GROUND TRUTH

A GEOLOGICAL SURVEY OF A LIFE

"Ground Truth is about the deep history of Oregon and the Pacific Northwest, a history built on geography rooted in geology. And, like geology, it is full of gems…"

—CHET ORLOFF, Executive Director Emeritus,
Oregon Historical Society

"Whether singing through bear country, uncovering a schoolyard toxic waste site, or confronting the loss of a beloved sister, Ruby McConnell is clear-eyed, kindhearted, and authoritative. Her stories, braiding together cultural, geological, and personal history, are by turns wry, gripping, and unflinchingly honest. And her writing is phenomenal."

—MARY DEMOCKER, 2019 Oregon Book Award Finalist and author of
The Parents' Guide to Climate Revolution

"…Timely, significant, and daring." —OREGON LITERARY ARTS

"In *Ground Truth*, Ruby McConnell weaves in the geologic history of the earth beneath our feet with contemplation of her life as a scientist and a woman. Each chapter is a complex and intriguing glimpse into both. She guides us beneath the surface of simmering volcanoes and heartbreak, an intriguing journey that leaves us wanting to know more."

—MARY EMERICK, former wil*
Geography of Water and *Fire in the Hear*

"I appreciate how [McConnell] has linked geological events and time scales into the story of her life span to date...the broader human life messages that [she] describes reach far beyond any arbitrary geographic boundaries."

—WENDELL "DUFF" DUFFIELD, U.S. Geological Survey (retired)

"Writer and geologist Ruby McConnell provides an unflinching and compelling embrace of the Pacific Northwest...a unique accounting of place and person that deftly balances personal memoir, earth science, and social history."

—ELLEN WATERSTON, author of *Walking the High Desert: Encounters with Rural America Along the Oregon Desert Trail*

GROUND TRUTH

A GEOLOGICAL
SURVEY OF A LIFE

RUBY MCCONNELL

GROUND TRUTH

A GEOLOGICAL
SURVEY OF A LIFE

RUBY MCCONNELL

OVERCUP PRESS
PORTLAND, OREGON

Published in 2020

Cover Art: Amy Ruppel
Cover Design: Jenny Kimura
Author Photo: Tracy Sydor
Book Design: Jenny Kimura

ISBN: 978-1-7326103-2-3

Library of Congress Control Number: 2019954226

Printed in Canada

Overcup Press
4207 SE Woodstock Blvd. #253
Portland, OR 97206

Overcupbooks.com
Rubymcconnell.com

*In loving memory of
Mary McConnell, chef.*

LAND ACKNOWLEDGMENT

THE AUTHOR RESPECTFULLY ACKNOWLEDGES THAT THE stories, histories, and landscapes portrayed in this book occur on the traditional homelands of people impacted by colonization, genocide, and displacement. Further, the author recognizes that a unique and enduring relationship exists between these people and their traditional territories in spite of their dislocation and the seizure of their land by the United States government and its people.

These people include:

The Kalapuya, Burns Paiute, Coos, Lower Umpqua, Siuslaw, Grand Ronde, Siletz, Umatilla, Warm Springs, Coquille, Cow Creek Band of Umpqua, Fort McDermitt Paiute and Shoshone, and the Klamath people and tribes of Oregon; the Chehalis, Colville, Cowlitz, Hoh, Jamestown S'Klallam, Kalispel, Lower Elwha Klallam, Lummi, Makah, Muckleshoot, Nisqually, Nooksack, Port Gamble S'Klallam, Puyallup, Quileute, Quinault, Samish, Sauk-Suiattle, Shoalwater Bay, Skokomish, Snoqualmie, Spokane, Squaxin Island, Stillaguamish, Suquamish, Swinomish, Tulalip, Upper Skagit, Yakama, Duwamish, Wanapum, and Chinook of Washington; the Bannock, Blackfeet, Coeur d'Alene,

Kootenai, Nex Perce, Northern Paiute, Palouse, Kalispel and Spokane Salish, and Shoshone tribes of Idaho; the Assini, Sioux, Blackfeet, Chippewa-Cre, Confederated Salish and Kootenai, Crow, Fort Belknap and Northern Cheyenne tribes, peoples, and communities of Montana; the Athabaskan and Alutiiq people of Alaska; and all other displaced peoples who call the Pacific Northwest home.

To learn more about Tribal Nations and the U.S. government visit the National Congress of American Indians at www.ncai.org.

To take action, write to your elected officials at www.usa.gov/elected-officials, expressing your expectation that the United States honor all treaties and agreements with the original peoples of the land.

TABLE OF CONTENTS

SUNDAY, MAY 18, 1980

"Vancouver, Vancouver, this is it."
—DAVE JOHNSTON

RAVEN-HAIRED, FAIR-SKINNED, QUICK TO LAUGHTER, AND filled with fire, I am my father's true Irish child in America, the youngest of three girls in a displaced clan, born of water and earth. As a "black Irish lass," my midnight hair casts me as a character of the ocean, a descendant of the selkies, the seal-women sirens known to sing sailors off their ships and out to sea. Legends say that any man who stole a selkies skin could keep her as his wife, marrying her for all eternity to the land. Those same ancient Celts believed that humans were formed of clay, shaped out of the very hills and bogs that nurtured and sustained them until life returned them to the ground. And it is true. The woman that I am today is the result of the processes and forces of this brutal and beautiful land to which I was born and am forever married to, a land of transformation and continuity in which I am destined to abide, endure, and ultimately return. To understand me is to understand the land from which I come.

The Pacific Northwest that you see today is the result of forty years of radical changes in the culture and economics of what was once a resource-extraction and agriculture-driven region. They are changes so fundamental in nature and scope that they effectively erased the region's history of pioneerism and work-a-day libertarianism. Changes that, for those of us from this place, will always be marked by the cataclysmic eruption of Mount St. Helens on May 18, 1980. On that day, I was just two years old. Everything that I have stood direct witness to since, everything I know about this place, happened after we watched the mountain crumble. The Pacific Northwest I was born to was a region digging out.

For a short time in the late 1970s and early '80s, my family lived in Seattle in a hilly, mostly residential neighborhood northwest of the city called Magnolia. Today, Magnolia is a reasonable approximation of a stereotypical Pacific Northwest urban neighborhood. It remains residential, with a small commercial district known as "the village" at its center. To the west is Seattle's largest public parkland, Discovery Park, home to the oldest lighthouse on Puget Sound. Smith Cove, at the base of the bluff, is a major port for the cruise ship industry. There is a sense of growth, prosperity, and progress created by new construction, businesses, and the accoutrements of the booming tech industry. It is populated almost entirely by white, middle-class, moderate liberals.

Like many things in the Pacific Northwest, though, the Magnolia neighborhood is not what it appears. Its namesake trees are really Madrones. Discovery Park is a former military installation. The lighthouse is one of the City of Seattle's primary sewage disposal points, which until the 1990s discharged largely untreated sewage out a twelve-foot-diameter pipe directly into the sound. Beneath the new development, the tech, the hipsters, and even the pioneer story lies a long history of racism, exploitation, thuggery, and environmental tragedy that gives Magnolia, and the entire region, a dualistic nature that allows both *Twin Peaks* and *Portlandia* to be true. The broad strokes of Magnolia's history also mirror those of the rest of the region, which was slow to develop in the first place because of remoteness and rugged terrain, held largely inaccessible by mountain ranges and what many regard as insufferable weather. A peninsula, Magnolia is bounded by water on all but three sides and only accessible by a series of bridges. It is the westernmost point in Seattle and the original home of the Duwamish people who are thought to have occupied the area for close to four thousand years. It was first claimed by white settlers in the mid-1800s when Henry A. Smith arrived and purchased more than one thousand acres of land around the cove in anticipation of the arrival of the railroad. It didn't happen for nearly forty years. Instead, shipping by water became the essential means of transport for the area—a logical choice for a city surrounded by the lakes and inlets created by the massive lobes of a continental glacier in the last ice age, nearly ten thousand years ago.

Later, the Magnolia peninsula served as the primary westward garrison for the U.S. military, Fort Lawton. Industry followed suit. By the end of World War I, close to twenty percent of U.S. warship tonnage came out of Seattle. By the World War II, twenty thousand soldiers were stationed at Fort Lawton, and a company called Boeing had begun to manufacture warplanes. In the mid-twentieth century, Washington State as a whole invested heavily in military, shipping, and manufacturing, as the Korean and Vietnam wars helped propel the area economically. The associated massive population growth (about twenty percent a decade) created a housing boom, which was fueled by the easy availability of cheap lumber and the willingness of local and state government to spend big on infrastructure. All the while, more and more of its workforce was folded into Boeing. Seattle by the 1960s was a booming, prosperous, one-company town.

The city celebrated its good fortune by hosting the 1962 World's Fair. Its iconic Space Needle was constructed specifically to cast Seattle as a modern, jet-set city. At the time, it was the tallest structure west of the Mississippi River, standing out against the city's mountain-dominated skyline: the crumpled ridges of the Olympic Mountains to the west and the Cascade Mountain Range's isolated volcanic peaks, most notably Mount St. Helens, to the east and south.

Eight years later, Boeing lost its government contracts and cut more than half of its eighty thousand workforce, enough to sink the single-industry area into a sustained recession. The OPEC oil embargo in 1973 didn't help. By 1974 local butchers were selling horsemeat and ground buffalo and Seattle's climbing suicide rate had spurred the installation of safety nets on the Space Needle. By 1980 interest rates were at twenty percent and unemployment was at eight percent.

"Will the last person leaving Seattle please turn out the lights," a phrase appearing on a billboard visible to all arrivals at the Sea-Tac airport and immortalized in a Waylon Jennings song of the same name, became Seattle's unofficial motto.

Such was the world into which I arrived in 1978, though I have few memories of our time in Magnolia. Like any distant memories, and particularly those from early childhood, they exist as isolated bright scenes emerging out of dark space. Some are composites, fables faded and supplanted by the memories of others, gleaned from the stories we tell and the migration of the truth with repetition. For instance, I have distinct memories of the giant octopus in the Seattle Aquarium that my mother would take me down to see nearly every day after dropping my sister off at school, the staff laughing as my legs kicked in excitement at (or perhaps in concert with) its massive tentacles. Or this, a vivid Easter Sunday afternoon egg hunt in 1980, memorable for the way in which my older sister dominated the hunt, trailing a parade of younger children like ducklings, and the presence of my Irish relatives. They had arrived en masse—cousins, aunties, uncles, and, most surprisingly of all, my tiny granny, all visiting the United States for the first time.

Pictures from the visit show them sitting in the raised yard, supported by fitted stone walls, surrounded by rhododendrons, azaleas, and Seattle's ubiquitous blackberry brambles, or smiling and sunburnt in the unexpected spring sunshine, standing at the edge of the Sound, squinting at the long, high ridges of the Olympics and holding me in turns at the top of the Space Needle, Mount St. Helens peak in the background.

Easter fell early that year, in the first week of April. The view of Mount St. Helens peak from my parent's yard on that day would have been different than perhaps at any other time in human or geologic history. Throughout the spring a series of small eruptions, bright white puffs of steam and ash, had dotted the sky over the mountain. A 1,500-foot-wide crater had opened at the summit. By Easter, the once nearly symmetrical cone had become bloated and distended with a great bulge on the mountain's northern flank that was growing at a rate of over fifteen feet per day. The mountain rumbled with moving magma. Scientists warned of "the big one," and in April drew a thick red evacuation line around the volcano on their maps.

For most of the people in the region, the impending eruption was little more than a curiosity. News of the volcano, at that time still privately owned by the Burlington Northern Railroad, which had received the land as part of a 47-million-acre grant in 1864 as trade for the construction of the Transcontinental Railroad, rarely made the front page of the paper. By May 15, just three days before the cataclysmic failure, *The Seattle Times* failed to mention it at all. It was no wonder. The major impact of the mountains growling had thus far been to force hunters, hikers, and a spare few property owners out of the area.

In contrast to the purpose of the evacuation order, the lengthy period of uncertainty had spurred a cottage industry- volcano tourism. The public, eager to see a live eruption, flocked to the mountain in spite of official warnings, encouraged by locals selling Mount St. Helens hats, T-shirts, and other doodads. Distances in the Northwest are big. Mount St. Helens is nearly two hundred miles from Seattle and surrounded by hundreds of square miles of undeveloped land, now a patchwork of BLM, Forest Service, and private timberlands but still held by the railroads in 1980. Geologists projected that even a major eruption would likely only directly impact these areas and the rivers leading off the peak. The largest of these, the Toutle River, drained south through an expanse of agricultural and forestland along the newly constructed I-5 corridor, still the region's only major freeway. For these reasons, perceived risk to human life and property was considered low, especially in Seattle, whose industry—what remained of it—would surely keep churning.

By Saturday, May 17, the *Times* was leading with a story about local residents storming the state roadblock by caravan, insisting that they be allowed back onto their properties "Come hell or high water." The state backed down.

At seven a.m. on May 18, 1980, though, life in the Pacific Northwest was still unfolding in pretty much the same way it had most of the twentieth century, the business of resource extraction abiding in spite of the economic slump. Even on a Sunday, log trucks rolled down mountain roads to load trains with lumber destined to become suburban tract homes across the United States. Shipyards hummed with passenger ferries and trawlers still full with salmon and crab. Children watched cartoons while parents slept in or enjoyed a second cup of coffee as they read the paper, which was dominated by distant crises in Iran, Florida, and China:

"New U.N. Effort to Free Hostages"

"3 Die in Miami Rioting After Ex-policemen are Acquitted"

"China Launches ICBM into Pacific"

On the mountain, only a few people remained in the evacuation zone, among them a stubborn old man named Harry Truman who refused to leave his lodge, a few curious hikers, Dave Crockett, a local reporter who had stayed the night in his car to cover the story, and Dave Johnston, a graduate student at the University of Washington working as a field scientist at the observation station located closest to the newly formed steaming crater. From his ridgetop perch, Johnston had a direct view of the mountain's bulging northern flank.

At 8:32 a.m. an earthquake shook the mountain. Johnston radioed the Forest Service headquarters in Washington with "Vancouver, Vancouver, this is it."

On the mountain, all hell broke loose.

The bulge rumpled and fell, propelled by a lateral blast, moving in excess of three hundred miles an hour, of ash, pumice, and gases that collapsed into pyroclastic flows—super-critical fluids that channeled down riverbeds melting the underlying glaciers and snowpack as they went. Fish jumped from the boiling rivers to the banks. It took less than forty seconds for the entire north flank of the volcano to fail, taking with it the top 1,300 feet of the mountain, Dave Johnston, Harry Truman, and fifty-six other people. Two hundred and thirty-two square miles of prime timberland was felled like matchsticks. The resulting vertical plume, degassed like a well-shaken pop can, erupted continuously for more than eight hours, raining burning ash, pumice, and chunks of ice on the entire state and spreading measurable amounts of ash all the way to the east coast.

Some survived the initial cataclysmic event only to struggle for hours in twelve-foot-high mudflows and heavy air fall, super-heated ash, pumice, and rock fragments that fell from the spreading eruption column. Crockett continued to film as long as he could, first shooting the sky as it filled with churning clouds of pyroclastic material, then pointing the camera down toward his feet, as he began to run. One man outran the eruption in an old pickup truck. Two more rode logs down the swollen, churning river to safety. A woman camping more than thirty miles from the peak was dug out of neck-high ash and mud by passers-by.

In Seattle planes were grounded and flights diverted. Traffic stopped. Streetlights turned on at noon. People wore gas masks and scarves over their faces. The city stood still and watched first as the mountain fell away, and then as the ash obscured the sun. The Pacific Northwest went gray.

In Portland my mother held me in her arms and watched the plume rise from our front yard. My father was fishing on the Clackamas River just south of Portland. He made his way home under darkened skies, stopping at a local sundry store for a pair of pantyhose to cover his exhaust pipe so that ash wouldn't choke the engine.

IN THE EARLY DAYS OF MY CAREER AS A GEOLOGIST, I STUDIED THE volcanic processes that led up to the eruption on May 18. The goal was to develop a predictive model, one that could be used on other, similar volcanoes or on Mount St. Helens, if it happened again.

Much of geology is guesswork. Many of the most important processes occur over timescales far beyond the human life span or even the course of human existence. And much of what we study is beneath the earth, forcing us to constantly extrapolate and imagine. Eyewitness accounts are rarely much better. In the U.S. Geologic Survey report on the eruptive events at Mount St. Helens in 1980, eyewitness accounts are treated with caution. The geologists who conducted the interviews noted that "two individuals recalling the same phenomena may differ markedly in their perceptions as well as their descriptions," and that quantitative data such as time intervals and distances provided by such sources are to be considered estimates as they are "often based on observations made under extremely stressful conditions."

Volcanoes in particular, are tricky. By nature, they are hot, fast, and, as in the case of Mount St. Helens, operating on such large physical scale as to force observation and data collection from a distance. My model was based in large part on remote sensing, aerial images, gas and seismic data, all collected at some distance from the actual events, limited by risk and probabilities. Geologists try to combat these limitations with ground truth, the process of going back and verifying observations and accounts made over distance and time, in the field. Without *ground truth*, all interpretations of the conditions and forces leading up to events are hypothetical, a story that we tell ourselves.

Before the scientific method and the reckoning of ground truth, there were other eruptions and other kinds of interpretations. The original inhabitants of the region might have attributed the eruption to the will of the gods, much as they did with the 7700 BC eruption of Mount Mazama, which collapsed under its own weight to form Crater Lake.

In my eruptive model, the story I told about the mountain, I posited that pressure built up and diffused over time as magma rose from the deep chamber to shallow levels within the volcano's conduit. With each rise, gases escaped out cracks and fissures, causing the magma to cool and begin to crystallize and clogging the escape of new volatiles. Eventually, a plug formed and pressure began to build, resulting in the formation of an unstable bulge, doomed to failure.

Perhaps something similar had taken place in Seattle.

From the distance of more than four decades it is clear to

see that the eruption's impacts to the region's industries—fisheries, timber, and transportation—totaling billions of dollars in damage and destruction, were only harbingers of what was to come as environmentalism, technology, and a persistent economic slump would encroach economically, drowning the pioneer culture with them.

Like Mount St. Helens, the region's economy, dependent for so long on the extraction of now-dwindling resources, would collapse fast. Nineteen eighty would be the last year more residents were born in Washington than in any of the other forty-nine states. Change would come in intangible form. Just one year before the eruption, Bill Gates and Paul Allen had signed a contract with IBM, sealing the fate of the region to tech and paving the way for Jeff Bezos and his mid-nineties startup, a company called Amazon, to change the face of Seattle forever. But before the region would dig out from the ash, before new saplings emerged and understory filled in, before Magnolia and all those neighborhoods like it would see revitalization, it would have to dig out of its economic and environmental crises.

My understanding of the region that I was born to is much like my understanding of Mount St. Helens, based on observations and measurements made from a distance. But one thing is clear: The forces and situations that have shaped my life and those of everyone in the region are knitted to the land, to forces both in and out of our control. In the Pacific Northwest, the land is always the thing. No exertion of human will or industry will take away its precedence. The course of the population's lives are inexorably linked to the whims of an untamable land. A dome still simmers in the crater of

Mount St. Helens. Trees still fall by the thousands, but now in the slow, noisy progression of industry doomed by greed and willful lack of insight.

What follows are fragments, field notes from those portions of this land that I have attempted to ground-truth for myself.

ACCRETED TERRANES

TERRANE: *A fault-bounded area
or region with a distinctive stratigraphy,
structure, and geological history*
—OXFORD ENGLISH DICTIONARY

THE PACIFIC NORTHWEST AS A GEOLOGIC REGION reaches from the western edge of Idaho and Montana to the Pacific Ocean, and from the Sierra Nevada mountains of California north, into Canada. Historically isolated and sparsely populated, the Pacific Northwest has in recent years seen some of the fastest rates of population growth in the country. This rise in population is not a reflection of internal dynamics but the result of a dogged migration, the region's first Great Migration of the twenty-first century, comprised of people fleeing climate change and the Great Recession, looking for a progressive paradise, and knowing little of the land in which they choose to settle.

It is easy, as one of these new arrivals, to foggily conceptualize the region linearly, as a series of north-south trending stripes beginning west at the Pacific Ocean and progressing east through coast ranges, inland valleys, the Cascades, and the vast high deserts and grasslands that lie to the east.

Reinforcing this north-south ethos is the I-5 corridor, which runs the length of the region south to north and along which the vast majority of the population lives. It is flat and straight, wedged between two mountain ranges of different origins, the wrinkled ridges of the Coast and Olympic Ranges versus the dotted line of symmetrical Cascade volcanoes. That nearly every city of any notable size is located along this corridor only serves to enhance the north-south perception.

Perhaps for this reason, a large portion of the population remains unaware of the majority of the landscape in which they live, preferring instead to stay within their stripes. Tied to the I-5, many will rarely if ever step forth into the arid grasslands, high deserts, and metamorphic ridges that make up the eastern two thirds of the land. It is a no-man's-land, home to the largest blank spot in the continental United States when viewed from space at night, Hells canyon, Chief Joseph's Trail of Tears, and the oldest rocks in the region, dating back hundreds of millions of years.

The Pacific Northwest has always been a land of migrations and immigrations. Every year, and for many years before the arrival of humans, thousands of gray whales migrate past the coastline, making their way from the Bering and Chukchi Seas near Alaska to Baja California to mate and give birth. Inland, Canada geese, whose numbers dwindled to twenty-five thousand in the 1980s, mimic the path the of whales. Hundreds of thousands of them now follow the Pacific Flyway annually to Oregon and California for winter.

The first people to occupy the region are thought to have arrived from Siberia via an ice bridge that formed across the Bering Sea during the last ice age ten thousand to fifteen thousand years ago. For thousands of years after that, indigenous peoples, the Nez Pierce, Chinook, and Salish, among others, occupied the territory, establishing seaside communities or migrating seasonally to take advantage of salmon runs, hunts, and valley blooms.

Exploration of the territory by Europeans began in the 1700s as businessmen and colonial governments became interested in new trade passages and easily extracted resources, but it would be nearly a hundred years before the region was considered accessible. Finally, in the mid-1800s white settlers from the east, more than fifty-three thousand of them, and Chinese laborers from the west converged at the end of the two-thousand-mile Oregon Trail. Soon, rail and shipping routes were established that carried lumber, gold and other minerals, beaver pelts, and salmon away faster than people in. Later in the century, racist policies like the Chinese Exclusion Act of 1882 pushed out this population and led to the arrival of a wave of new laborers, Japanese immigrants, whose later interment during World War II would forever scar the region. By the 1960s the Northwest had established itself as the heart of the back-to-the-land movement, drawing thousands of Jesus People, New Agers, and Commune builders to its cities and secluded valleys.

These migrations, in the context of the entire history of the region, are largely insignificant. From the perspective of the land, humans are new arrivals, clinging to the edge rocks of the continent like limpets in a tide pool. The geologic history of the Pacific Northwest spans 600 million years. For much of that time the region has been an active margin, a meeting place where oceanic and continental crust collide and exotic terranes are imported across oceans and millennia. In such a collision the relatively light, less-dense continental crust always wins, overriding the oceanic crust which is thrust downward, into the heated interior of the earth in a process called *subduction*. The geology of the Northwest, the land itself, is a product of the horizontal forces of this collision, which runs in direct contrast to our vertical self-perception. We are, in fact, a region of lateral forces applied at right angles to how the modern population has arranged itself.

When the Precambrian era was giving way to the Paleozoic, 550 million years ago, the western edge of the North American continent was a passive margin, tectonically inactive, terminating at the current western border of Idaho. To the west, where Oregon and Washington now are, was shallow sea. The transition into an active margin began when the adjacent oceanic plate began to move east, bringing with it seafloor sediments and exotic landforms from the far side of the Pacific. When the plates collided, the weight of the oceanic plate plunged it beneath the continent and the seafloor began to subduct beneath North America. As subduction progressed, the exotic terranes, bits of continents and seamounts from the far west accreted, coming together with the edge of the continent. These terranes, more than fifty in total, begin at the eastern edge of Washington and extend west to the current coastline where they can be seen at the surface as the more than three hundred San Juan islands that speckle Puget Sound. For the most part though, they are basement rocks, sitting, invisible and unexposed, at the bottom of younger deposits, the sandy gravels of the Missoula flood and pumice and ash from Cascade volcanoes like Mount St. Helens.

Like the tectonic plates, my family was carried here by forces larger than ourselves, arriving from both directions.

My mother came from the west, Australia, a continent of koalas and kangaroos, and deserts and didgeridoos that has remained geologically stable for the better part of the last 200 million years. She is the daughter of an American soldier and an Australian teenager.

During World War II, the U.S. Army had a program called "Operation War Bride" that promised free passage to the United States to the wives and children of GIs abroad. Since the Australian government also had a campaign encouraging young women to entertain the more than one million U.S troops that would pass through the country during the war to boost morale, marriages were common. As approval from the military was prerequisite for immigration, the newlyweds were often then encouraged to have a child on the way to expedite the process. Still, many of these women waited years for passage.

In Brisbane, Australia in 1943, my Nana was seventeen and did not go to high school. Her mother was a cellist, her father the kind of gambler that made her early in life accustomed to going to bed in a mansion and waking up in the public park, her father having gambled everything away and moved her out while she slept. At seventeen, she left home. With few prospects and no resources, she worked days on the base and danced with GIs in the evening, boosting morale in the clubs. At night, she would sneak back onto the base to sleep in one of the empty jeeps parked in the army's vast lot.

My grandfather, twenty-two, was an infantry private stationed in Brisbane at the United States South West Pacific Headquarters to recover from malaria, which he had contracted while fighting in New Guinea. He had seen my Nana sleeping in the jeeps. After many nights of watching her, he went over, opened the door, and woke her up. He told her that he recognized her, that she was one of the girls that danced, and that he had seen her sleeping in these trucks for some time. She told him that she recognized him too—one of the guys who drank and played cards for money during the day. They talked that night and other nights. He started to seek her out at the jeeps and at the club, and he asked her out, several times, before she finally said yes. They arranged to meet at the movies, a common date back then, as GIs weren't allowed into theaters without an Australian escort. That night, she went to the theater and stood across the street and watched him wait for her for a long time. Eventually, she told my mother years later, she thought, *Do I really want to do this?* and then just walked across the street.

My mother was born the following December. Three years later, she and Nana boarded a ship with hundreds of other women and children, on a multiweek journey that would deliver them to unknown family in the Pacific Northwest. Tens of thousands of Australian war brides came over this way in the years following the war, landing like geese along the shores of the west coast.

My father came from the east. Born in Northern Ireland in 1939, the oldest of four children. His father died when he was nine, leaving his mother and her three unmarried sisters, his, "maiden aunts," to raise them in the Catholic ghetto of Derry during the war. By the time he was a young teen it became clear that he would have only two choices in life—to be a farmer or a Catholic priest. He attended All Hallows Seminary in Dublin, graduated at eighteen, and ten years later, in 1965, was told he'd be moving to the United States.

His immigration was a cog in a larger cultural, religious, and political movement that had been taking place over hundreds of years. Ireland began sending Protestant priests to what would become the United States in the 1500s as part of King Henry's reformation, but after the potato famine in the mid-nineteenth century, the Irish Catholic Church was revitalized and began producing priests-for-export to English-speaking countries. When Irish independence was achieved in 1921 the protestant persecution of Irish Catholics in the North spurred a Devotional Revolution that would generate several generations of socially minded and fiercely nationalist young Catholic priests. Many of them, ill prepared for a life outside the seminary, would arrive in the United States in the mid-twentieth century. Unaccepted by the well-established Irish-Protestant diaspora which rejected Irish independence, these young priests often aligned with the newly radicalized left. Upon his arrival in Oregon, with its running water, radios and television, and comparatively spicy food, my father decided that he had found paradise in this familiarly green but still uncharted territory. Guided by Catholic principles of peace, he began registering conscientious objectors to, and marching against, the Vietnam War, a streak of liberalism that found him soon relocated to an older, more conservative parish in Portland where the Archbishop thought he would be kept out of trouble. A few months later, he met my mother at a church potluck.

When they married two years later in 1970, most of Oregon and Washington as we know it now did not exist. The Northwest was an inland sea of a different kind, separated from the rest of the United States by distance and a cultural void, a land of tenuous future. The I-5 was yet to be constructed, Civil

Rights were just beginning to be implemented, Roe v. Wade did not exist, skyscrapers were few, and the old guard, timber, railroads, and fisheries still dominated the economy. It was a period of cultural transition and uncertainty that offered limited opportunity to new arrivals. For them, like many immigrants, the formation of new family annealed their connection to this place more than anything else, and even that, like the land they built their new lives upon, was patchworked.

There were three of us girls. We were stair-stepped, with more than half a decade separating us. We each had raven-black hair, but that was about where the similarities ended. I arrived last, a surprise. But Mary, the oldest, she was chosen.

Before they even met or married, my mother, a white, underemployed single woman living at home with her parents, had adopted a three-day-old olive-skinned baby, Mary, my oldest sister. We would never know anything of who or where she came from, just that she had arrived first, before even Dad, our family's most exotic terrane. Together, they raised us all as Irish-Americans, speaking scraps of Irish in the house, making soda bread with a Sunday morning fry, and taking us to dance reels and jigs at the Irish céilís held on a Friday night in the Police Union building in downtown Portland.

Neither my father nor my Nana ever lost their accents. Nana never even became a citizen, remaining on a green card for forty years and proudly announcing to anyone that asked that she simply didn't want to be an American citizen. Homesick her whole life, she finally persuaded my grandfather to return to Australia in their seventies. Dad, after years of working to Americanize, reclaimed his heritage during The Troubles of the 1980s when Ireland forced its way onto the international stage.

Now in his seventies, he seems more Irish than in my youth and, like my Nana, yearns for his homeland in a deep way that drives him to spend weeks and months a year in the narrow streets and rolling hills of his childhood.

Like people, terrane rocks retain the characteristics of their previous settings. Which is why, in the Pacific Northwest, you can find seashells on mountaintops.

Millions of years ago, in the Mesozoic, migrating micro-continents and seamounts finally displaced the shallow inland seas that were Oregon and Washington. The first and largest of these, the Intermontane Terrane Complex, amalgamated at sea long before crashing sidelong into the continent to Washington and British Columbia. Others, the Blue Mountains and Wallowas, rafted to the continent in isolation. They did so carrying with them preserved remnants of the tropical locales from which they came: coral reefs, sandstones, and mudstones rich with fossils of mollusks, sponges, sea urchins, and early fish. Each with its own assemblage of plants and animals imprinted in its sediments, many of them retaining, startlingly, the magnetic signature of their place of formation—their volcanic rocks indicating a different latitude and longitude at the time of their eruption and cooling, the magnetic minerals forever pointing to a false north.

This process, this annexation, the integration and annealing of a new terrane to a continent, is accomplished by the application of heat and pressure, locally, and as regional forces. The process is tortuous, as one landmass is forcibly joined with another. The resulting metamorphic rocks can be so sheared, faulted, and convolute with folds as to challenge the even language of geoscience, folds turning to thrust folds turning to overthrust folds in description. It is common in annexation for terranes to be compressed to half their original width. With time, millions of years, magma is intruded into the joints and fissures that result from the collision. With more time, the magma slowly cools and crystallizes into granite. The cooling releases heat into the surrounding country rock, causing it to recrystallize in a process called contact metamorphism. These cooled igneous veins, visible as vast bands cutting across the strata, act as sutures, annealing the terrain to the continent. The current suture point runs north-south along the western base of the Cascade Range, roughly the path of the I-5 corridor.

Perhaps our new arrivals then, have a point in their linear view, their perception of us as so many parallel-aligned minerals. Perhaps from a vantage at some distance it is easier to see the bands and folds of humanity as they have accreted to this terrain, tied, as we are to water and resources, mountains, and coastline. If this is true, what vestigial sediments have we brought with us? What fossils will be found from our time on the edge rocks? And how can we know what to think of this new migration, this new influx of population that is being thrust upon us?

The inevitable consequence of accretion is metamorphosis. Likewise, the true inner workings of this place as we know it, from our human perspective in this point of time, are processes of transformation driven by heat and pressure, applied directly and via forces from afar. The result is not a melting pot but a quilt, each patch retaining its unique character, still essential to the whole.

AT THE COUNTING
WINDOW

F OR THREE MONTHS NOW I HAVE BEEN THINKING ABOUT the salmon. It started, somehow, in thinking about home, or nesting, or love. Or perhaps it was the receipt of bad news, an old friend, long sick, passing on; the anniversary of another friend's death. I had just returned from long travel. At first it was just the word, *salmon*, that came to mind, mostly in times of melancholy. Then it was images, the rush of water, silver-green streaks of light infused with sadness and joy. Then memories: my mother with a camera urging me as a very small child to stand still as my father, just home from fishing, dangles a Chinook next to me for scale. In quiet times, I have begun to watch them jump upstream in my mind's eye, reaching, over and over again. I have been under some stress.

Now, in the solitude of winter, I revel in these moments with the salmon. I find the photo in an old album. In it, my father grins broadly next to us as I carefully consider my cohort. I sit on the porch of my house in the trees, close my eyes and let the salmon jump. I take deep breaths. For a moment, my mind shifts to the holidays, freshly over. I think I should have brought a cup of tea out with me. Or some whiskey. I wonder how many eggs each salmon mother lays, if she will ever know how many children she has. I return to the stream. I eat a stale cookie and think about Christmas. The thought that my parents are getting old emerges. I'm tired. I must be getting old too. I think about a recent visit with a good friend who told me that we are all responsible for keeping ourselves above the water.

I am an Oregon native; like the wild salmon of the Columbia River run the streams of my youth come from a single source descending first to the Willamette, then the Columbia, then out to the Pacific Ocean. Like all Oregon children, the story of the return of the wild salmon to their native headwaters year after year to spawn and die has been pressed upon me like the grooves of a well-worn album. How they are first shunted out down rapids and into the salty shock of the open ocean where they gorge themselves on baitfish. How they return, by scent, to the river's mouth and begin the long and arduous climb to their spawning grounds, the grounds of their parents, and their parents before them. How there, after spawning, they die and nourish the waters for the next generation.

The stories describe the salmon as prodigal sons, overcoming all obstacles, debris, dams, fishing nets, and sea lions. They tell of the salmon as Pacific gold, the lifeblood of the Northwest people and the keepers of traditions. In the best of these stories, the bravest of the warriors are the salmon of the Columbia River run: my run. These are the salmon that once ran so thick the fisherman could walk across their backs to make the river crossing. These are the salmon forever immortalized in pictures of the Native fishermen, on wooden platforms, salmon jumping the falls over their heads. These are the salmon I see now behind closed eyes. How do they know what to do? I wonder. Why do they fight so hard?

I drift to a memory of myself as a child, grasping my father's hand in a darkened room while the salmon run past a picture window inside the Bonneville Dam. The room is oddly silent save for the clicking finger of the fish counter seated slumped on a stool next to the glass. At first, disoriented, I am not sure what we are watching, how such a window can exist inside the water. I wonder at what kind of place this is. The drive up the gorge had been long; the outside air was sharply cold and heavy with moisture. Somewhere on the old highway we had to slow for a bird of prey standing in the road. The dark green of the trees pushed against the white-gray of low-slung clouds. When we arrived, the dam stuck out in awkward contrast, a massive, colorless slab of concrete. Inside, the weight of the dam itself, and the energy from the powerful turbines has overheated the air. At the counting window the salmon are in a thick frenzy against the glass, pushing forward in a mass of frantic energy. My stomach grumbles and I wonder how long we will be at the window.

My stomach did not know that it was irreverent to complain at such a time. At six, I did not know that some of those fish would have waited for more than a day at the base of the dam before entering the ladders. I did not know that some of them would travel nearly nine hundred miles and gain some five thousand feet of elevation before reaching their destinations. I could not grasp the enormity of the struggle I witnessed at the counting window, the frequency with which they would lose the battle, or the tenacity required to forge ahead. I did not know that I, unlike the fish, was on my way out, not back. I was only just making my way downstream and who knew how far adrift I might be cast?

In school, we would learn about salmon. So much so that the salmon would become the connecting thread between classes. We would learn about Lewis and Clark, their journey and deliverance to the land of trees and salmon. And about the Native peoples who honored the salmon and other spirits. The salmon would be how we were taught about resources and our fragile interdependencies. At home we were fed great thick steaks cut from the fat salmon our fathers carried home in worn red coolers. Everywhere else the salmon populated public spaces in murals, sculpture, wallpaper, greeting cards, and coasters. In these images the salmon are always shown swimming upstream, moving with power and purpose, often depicted not as creatures of the water at all but as some kind of flying beings. At six, I had not yet learned these things, but I was old enough to know that salmon were important. For the moment, I kept still, quietly eying the scene behind the glass.

A few weeks before the trip to the dam I had received, in lieu of the more common gold cross or family rosary, a fishing rod for my first communion. I remember my father, in response to the politely stunned silence that followed its unveiling, listing in all seriousness the levels of symbolism: coming of age, returning home, loaves and fishes, the fish as a symbol of peace and abundance. "Teach a man to fish," he said, as my mom handed me a small green tackle box with my name carefully spelled out across it in stickers.

And teach me to fish he did. In the coming years I would spend long lazy days along rivers and lakes, rod in hand.

Listening, sometimes, to my dad tell me stories from his childhood, more often sitting in relative silence, waiting for the line to tug. Which it rarely did. The truth is, despite one's experience and previous good luck, salmon will earn their reputation as both resilient and elusive. A twelve-year-old with a pint-sized rod proves to be a match for only the most unlucky. The salmon are fickle as well. I remember a day we sat for hours without so much as a bite only to have a man arrive and catch three fish in twenty minutes not a hundred feet downstream of us. "Murphy's Law," Dad would say, shaking his head.

Back in the belly of the dam, I lean my face in to the glass and watch while inches away a large sockeye fights his way upstream. I wrinkle my six-year-old nose at him. It looks like it requires an enormous amount of effort. He calmly stares back at me through the glass, his one, unblinking eye the only part of his body not moving. Temporarily losing the battle, he shoots backwards and out of sight only to emerge again, regaining his previous position for a moment, then pushing forward. I look up at my dad; he releases a deep breath and nods at me as if to acknowledge some kind of understanding with the fish. We pause a moment, looking at one another. I turn back to the window. Silently, we watch the salmon swim.

I have learned other salmon stories since that day in the dam. Stories about logging and dairies and sediment content; stories about tribal lands and fishing rights and casinos; modern stories, in which we paint ourselves as both the downfall and salvation of the salmon. In these stories the salmon are not wild, but of our own creation. One day I stop at a hatchery on my way out to the coast. I stand, looking down through a metal grate at what must be thousands of small, test-tube salmon in concrete tubs. It is raining and hard to distinguish the hatchlings from the moving water. I think about the day at the counting window. They swim in a tight school, full of youthful energy. I wonder if they know they are not wild.

On my porch, looking at the trees, the story I want to tell myself is the one with the happy ending. I want to tell myself that I have arrived home, safe and triumphant, having faced my obstacles and endured the uphill climb. I want to celebrate my survival to this point and begin to look back with fondness over my years. I want to know the danger is behind me. But I know that for the salmon, at least, the journey takes a lifetime; and I am not yet old. Still, I think, I am not so young that I do not have an understanding of my situation. I have just as much chance to make it past the counting window as anyone else. I close my eyes and watch the salmon jump. Upstream, always.

SWEET MILK, THE HOUSE IN MORNING

N EARLY MORNING I STAND AT THE KITCHEN WINDOW AND watch the fog lift out of the trees. I woke first, as always, though today roused early by the sound of the rain pounding on the roof, the gurgle of water overflowing from a clogged gutter, and a deep, aching pain across my low back. In hazy solitude I rose, rubbed my back, and faced myself and the new day in the dim light of dawn; there was no fear of waking you, even the alarm does not stir you most days. Through the kitchen window a deer and her fawns stroll lazily past, nibbling the moss off fallen branches knocked loose in last night's storm. The early spring storm had rolled through the night carrying great gray thunderheads that filled to overflowing the soils already soaked from winter rains. The winds shook free the last of the season's loose limbs and upended the neighbor's newly purchased but yet to be planted spring starts. Unlike the tedious drizzles of winter, spring rains are temperamental and arrive in fits and bursts that shatter the tenuous calms of the first warm months and threaten the purchase of early growth. After a pregnant pause, the rain begins again.

Although known for rain, the Pacific Northwest is divided roughly in halves between the rain lands and the high desert. The Cascade Mountain Range, which runs the length of the region north to south, acts as a great catchment, trapping the clouds in the western lowlands until they release enough of their watery load to make the climb over the peaks to the high plateau lands beyond. By the time they reach the far side they are all but rained out, leaving the east comparatively barren. In a normal year, portions of the high desert will receive more snow than rain, though scantly enough to support the thick stands of ponderosa pines that quickly give way to grasslands as you move east, away from the base of the range. In the west, the rain is the sweet mother's milk of survival for the densely packed understory and giant Douglas firs and hemlocks of the temperate forests. In the spring, the warm rains melt the winter's snow, swelling the rivers and filling wetlands with fresh water that will sustain them through summer. The spring rains also call the bloom, an embarrassing display of fertility that forces the deep green of the trees to play second fiddle, for a time, to the purple lupine that dot the roadsides, bright orange tiger lilies that hang like lanterns, and the gentle pink of wild roses.

This spring, even with the recent storms, we are behind. *Behind*, what people from the Northwest say instead of drought. In a place as wet as this, where average rainfall can be more than eighty inches a year, talk of drought seems like hyperbole. In these years the rains still come, the crops still grow, and we enjoy a long and lingering camping season. It is easy, in this land of abundant water, to forget the desert lands to the east and their fragile dependence on our leftover precipitation. Easy to forget the ranchers to the south that wait in vain for our rains to work their way through the subsurface and into the wells and channels that feed their herds. Easy too to forget the salmon runs and the dams that power our homes. We may stand and marvel at the exposed underbellies of our reservoirs, speckled with drowned trees or lament a bad ski year, but we are slow to raise an alarm over being a year or two behind. The Pacific Northwest as a land of water is so deeply knit into our understanding of the world that we are unable to grasp the consequences of being behind for another year. There is always time to catch up, we think. There will always be another rain.

I run the water in the sink until it warms and start to do the dishes from last night's dinner. My mind travels across a list of things to do. I too, am running behind on things this year, I think: late. In these early days of spring the air still bites with cold and the rain often falls as hail, having frozen at higher elevations or lands as ice, freezing to the chilled ground. Last night's storm ripped through the fragile early cherry blossoms, littering the ground with pink and white petals like festive confetti. Just as quickly as it began the rain eases and then disappears altogether. The clouds pass away into the hills and everything is illuminated in a wash of bright white light. The world holds its breath as the water continues to fall off trees and eaves. Then the birds begin to sing in uneven harmony, songbird twitter percussed with the harsh calls of ravens and the five-beat rhythm of a solitary crow.

I look down to discover my hands have turned red under the faucet. I have been so tired, unable to sleep well and waking early to drink peppermint tea, the only thing my stomach seems happy with these days. My low back aches with a dull and constant pressure. It worries me. I lost the baby on a morning just like this. That morning I had lingered in bed after a bad night's sleep, my back aching. You rose uncharacteristically before me, hurtling into your day in a flurry of shower and food and activity. I finally followed the trail of you left over from the previous night, up the stairs to send you off. You stopped, halfway out the door, pausing to read my face. "What's wrong?" you asked. "I don't know," I said. "I'm bleeding."

I make myself a cup of tea and look out at the rain-soaked trees and try to shake away the twinge I feel across my stomach. I remember how afraid we were that day, the way you held my hand, helpless, for hours. The months it took to recover, return to normal. I think about my mother, who taught me most of what I know about the rain. Or at least those things it could not teach me itself. My mother is a crafter and a poetess, a weaver of things and words. She makes baskets and pies and quilts. She writes deliberate and considered poems. In one, she attends a mass that is interrupted by a homeless man. She describes his struggle to express himself at the podium, the silence of the congregation, his urgency, the way his hands fluttered around his face, like the wings of a moth beating against the glass. Watching the rain streak down the windows I feel that way myself, struggling to emerge, my wings beating against the glass.

I remember her as I am now, wrapped in a caftan, sitting with a cup of tea in front of the picture window, contemplating the morning rain. I see her hands, delicate and sinuous, one with a simple gold band, the other with a bright green stone, resting quietly in her lap. In this memory she is telling me the story of myself, of how I came to be in the world. She explains that I arrived late to the family, fourteen years after my older sister. "You were not supposed to be able to be," she says. "But boy, did you show them. The doctors had told us that your sister would be the last, and besides, so much time had passed, I didn't seem to be the age for making babies anymore."

I consider the age for making babies. I am the age my mother was when she had me. I wonder how we decide that we are able to be. I do not remember deciding to be, though it seems I must have. Outside, the rain has changed to mist. Like the native Alaskans who have at least thirty words for snow, the people of the Northwest have at least as many words for rain. Functional words that acknowledge its subtleties and temperaments, and poetic words that describe the gentleness of coastal mists and morning drizzles as they stand in contrast to fierce fall downpours and winter's freezing sleet. My mother taught me most of my words for rain. She also taught me to use the rain as good reason for the making of home, a thick soup or loaf of bread or to tuck into a good book, or play a game of cards. She taught me that the rain can reflect the edges of your soul, that we too can shift in temperament and time. That the clouds can invite you to sit and have a good cry right along with them, and that, at times, it is important to do so. She knew how to live alongside the rain, not in spite of it like so many of the rest of us.

I think back to a gray fall day from my childhood. We were on our way home from school and work and my mother and I pulled up outside the house just as the skies opened in a heavy and drenching downpour. It was late November and already fiercely cold with the east wind out of the Columbia River Gorge. The rain thundered against the roof of the car. The streetlights were already on, though it was scarcely half past five. My mother looked at me, in just a turtleneck and jumper in spite of her reminder to bring my coat, then turned and contemplated the bags of groceries crowding the back seat. "Maybe we should just sit here and listen to the radio for a few minutes while it passes," she said. In this she was not being entirely hopeful, the weather in Portland moves fast, and often a fifteen-minute wait will turn a storm to sun, or at least a rain to drizzle. I considered the sky, blanketed in woolen gray and weighed the odds of it clearing against the sogging trek up the stairs to the front porch. Steam was still rising up from the heated hood and Patsy Cline was playing on the car's tape deck. "Okay," I said. She smiled at me, and settled back against the car door, humming quietly along.

Two minutes later, it was as though the gates of heaven themselves had opened and all the waters of eternity had been loosed upon us. Great blasts of wind-driven sheets of rain splashed across the dashboard like buckets of water being tossed on us. The car shook slightly with the impact. It was not clearing up. "Well," she said at last with a sigh. "I suppose we shall have to take what the rain gives us. Are you ready?"

I don't know, I think. I hear the shower start from downstairs. You are awake at last. I get up to finish the dishes and put the water on for coffee, then stand at the window and cross my hands over my swollen-only-to-me belly. Inside, the dull ache has turned into a sharp pain. I wonder if we decide not to be as well. This time, I am not so fearful. Outside the wind has picked up, bringing in a fresh storm. Fat drops begin to slap across the window. At least, I think, there will always be another rain.

THE VIEW FROM COUNCIL CREST

MARY MCCONNELL
1964–1996
A celebration of the life of Mary McConnell
will be held Saturday, April 28,
at The Newman Center, Portland, Oregon
Mary, a chef, is survived by her parents and two sisters

O N A GRAY NOVEMBER DAY, I SET OUT FROM THE HOME
of my parents on the flank of Mount Tabor in southeast Portland and headed west onto the flats towards
the Willamette River and out of town. Somewhere near my
old high school I took a detour, up the winding roads leading away from downtown, past the old-money houses of the
West Hills, over the Vista Bridge and up to Council Crest, a
concrete-topped grassy mound overlooking the city. From the
viewpoint looking east you can see almost all of Portland.

It is a view I had not seen in nearly two decades, in no small
part due to the influence of my personal history. Twenty years
ago, my sister Mary died on that hilltop and I have never had
the courage to return. Now though, I did. I told myself that it

was an excursion driven by an urge to test the fragile boundaries of my recovery from the loss, the extent to which I had been able to move on. But it was just as much about an aching need to reconnect with a sister whose last act was to drop me off at school at the base of the hill, calling out "I love you" as she pulled away from the curb. It was time for me to face the past.

The official history of Council Crest is varied. Local legend claims the lookout received its name after its early use as a meeting place for the local Native American tribal councils, but in truth the name refers to the Triennial Council of Congregational Churches which met there in 1898. Despite this principled beginning, the virtuous days of Council Crest were short-lived. By 1907 it was the site of an amusement park and dance hall open every night except Sunday and billed as "Pleasure Park—The Dreamland of the Pacific Northwest." Its popularity persisted for decades; a 1925 newspaper ad for the hall still encouraging Portlanders to "Get that healthy, happy Crest feeling!" Back then, Council Crest was the place to be, especially if you were looking for a good time.

Not a lot has changed over the years.

By the time I was in high school in the early 1990s, Council Crest was well known as a popular place for kids to party at night, away from the prying eyes of parents. The low, circular stone wall surrounding the lookout made a perfect place for us to hang out, drink beer, behave badly, and break rules, especially after dark. Most weekends and many weeknights, you could find at least one or another group of kids hanging around, smoking pot in the secluded corners and forested nooks that ring the viewpoint, playing music from their cars, and getting into harmless trouble.

We were not alone. Regardless of time or day, we had company, and not just joggers, bird watchers, and families. Mixed among us were couples playing out illicit affairs, prostitutes with their johns, and the proprietors and clients of Portland's robust and obvious drug trade. There were cars that sat idling until speeding off after a quick exchange, junkies napping on benches, the occasional needle. We never really thought about it. It seemed normal. The view from Council Crest seemed to beg intoxication.

At 1,073 feet above sea level, Council Crest is the highest point in Portland. It is a comparatively small park, occupying just forty-three acres of land, but it has persisted for over a century as one of the most popular green spaces in the city on the weight of the view alone. On a clear day when the sun is shining you have a 360-degree view of five Cascade peaks and more than three thousand square miles of Northwest territory. You can look down onto the crowded almost-skyscrapers of Portland's business district, across the river to the squat brown-box buildings of the eastside industrial area and the boutique and brewpub-lined blocks of Hawthorne Boulevard that lead up to Mount Tabor, the small volcano that sits squarely opposite Council Crest, my parent's house perched on one side, the reservoirs that hold the city's drinking water on the other. On days like that, Portland shines a glimmering green and the light reflects off the Willamette River towards Council Crest which rests above it like a pristine jewel.

It was on such a clear day, a sunny Easter-week day, that Mary dropped me off at school, drove up into the hills, wrote a suicide note, and overdosed on heroin. Her body was found two days later, on Sunday morning by a little girl on an Easter egg hunt. For us, my family, it was the end of a long personal tragedy of detox, treatment and rehabilitation programs, halfway houses, and arrests. For Portland, it was one more sad side effect of an unacknowledged underbelly: the tarnished reality of an old frontier town founded on fresh starts and the promise of prosperity.

Looking down from Council Crest it is hard to imagine; that Portland has a drug problem is, somehow, well hidden in plain sight. It is not a side of itself that Portland advertises. The image we would like to project is that of a hip, artsy, and educated metropolis in the trees, a clean and well-planned city inhabited by healthy, attractive people, known for their harmless eccentricities, eclectic restaurants, and taste for craft beers. But the drugs are here. You can see glimpses in crime statistics and as undercurrents of gentrification discussions, but rarely is it discussed with any frankness, and never with respect to the high-end habits of Portland's upper class.

And it does hide. When I walked through downtown to school every day, I was never approached by a drug dealer, but I remember being shocked one day when we were offered drugs three times in a single block of Broadway. Mary shrugged it off as dealers being able to spot an addict a mile away, but it opened my eyes. Apparently, in Portland, heroin was as easy to buy as lattes, if you knew where to look.

My sister, like most addicts, was part of two worlds, two Portlands. A well-known chef in the ever-trendy northwest neighborhood that sits at the base of the West Hills, she was at the center of the proto-hipster scene of the eighties and nineties. She described the time as being heady and laden with drugs; it was common for her to be tipped with cocaine on a Saturday night. A week after her arrival home from a job in New York to detox for the first time, I opened the paper to a column celebrating her return to the Portland restaurant scene. She was at the forefront of the celebrity-chef movement and the phone had been ringing for days with restaurants eager to offer her a job. In those conversations the reasons for her return and the conditions of her rehabilitation were never discussed.

Drugs and money in 1990s Portland ran downhill like water, toward the Willamette. Like Mary, most people in Portland didn't buy drugs on the street. Usually, she got her drugs at bars and restaurants through an open network of people in the scene fueled by salon-drug dealers based out of their houses in the hills. One cocaine ring operated for years, delivering drugs on demand out of expensive cars in the parking lot of a high-end grocery store before being busted in a sting operation. The farther you got from the hills, the less glamorous it became. When Mary was finally arrested for drugs, it was one mile away, on lower Burnside, one block from the river, long before the transformation of the area from an industrial wasteland into the upscale Pearl District. It was, and still is, where people go to hit bottom, when they have run out of fancy friends and money, their desperation making them the easiest pickings for law enforcement.

Mary was thirty years old when she died in the park, statistically average for a heroin user in Portland. And she wasn't alone. Heroin-related deaths in Portland rose six hundred percent in the '90s and have spiked to those levels again in the second decade of the 21st century. Part of this is due to geography. Black-tar heroin moves quickly up the I-5 corridor from Mexico, and Portland serves as a primary distribution point for the rest of the country. This matters to heroin addicts, because it makes for a lot of variation in potency. You never know what you are going to get, so it is easy to accidentally overdose on Portland heroin. For many addicts, like Mary, overdose is not accidental. Portland's suicide rate is among the highest in the nation.

A lot of those suicides happen in the West Hills, many of them from the Vista Bridge, which I crossed on my way to Council Crest. Perhaps this is because of the area's combination of beauty and solitude, the same reasons people go there to drink and get high in the first place. So, Council Crest closes at midnight, when a city worker arrives and shuts the big metal gate that swings across the access road. Ironically, it is exactly this security precaution that makes it such a great place to party. The closed road forces you to park in the neighborhood, blending your car into the mix on the surrounding residential streets, and arriving on foot makes you all but invisible once you walk past the gate. With the road blocked, the beat cops can't drive in either, and they are disinclined to walk the quarter mile uphill. Once the gate closes, there is no one to see you. And in Portland, what isn't seen, doesn't need to change.

On the day I finally return to Council Crest, the mountains are obscured and a bracing east wind drives the rain into the windshield. I arrive mid-morning on a weekday in the hopes of having the place to myself. When I first pull up there is one young family making an effort at a picnic. I walk the long way around the road and by the time I reach the summit the family has retreated to the car, the heat from their bodies already covering the windows with steam. The water tower looms above me. When I reach the viewing platform, I look down at the plaque pointing out the peaks. I stand with my feet pointing to Mount St. Helens and peer into the rain. I see nothing but a wall of gray clouds. I try to imagine the view as it was when Mary ended her life. I imagine it as breathtaking—the trees leafing out and trillium blooming along the trails. She would have been able to see the mountains. There must have been others in the park on that day, people walking dogs, joggers passing through. I wonder if they saw her there, if she was a part of their view from the crest, if to them she looked like a chef or a junkie. I imagine that either way, they were all minding their own business. In Portland, in the hills, it is assumed that people will mind their business.

I pace the lookout in a slow circle, framing the scene. To one side of the grassy rise is a bronze statue commissioned by the city in 1956, a scant ten years before Mary was born. I stand looking up at it from its base. It depicts a mother and child, suspended in a moment of joy and play, the mother floating the child in the air above her head, his arms and legs outstretched as if in flight. It reminds me intensely of innocence.

But everything at Council Crest has a dark history. In the 1980s, after thirty years of flying above the city, the statue was stolen, the mother's legs cut below the knees and the whole piece hauled off in a long and laborious late-night caper. No one saw a thing, and therefore there was no substantive search for the pair, or any effort to replace them. Ten years later, just after Mary's death, there was a dramatic rescue; the statue was discovered in the backyard of a Portland home during a drug raid and returned to the park. I inspect the mother's legs, looking for signs of her ordeal, but there is nothing, and no mention of the incident on the plaque, the past erased, washed clean. Like many things in Portland, you would never know to look at it.

Behind me, a car with tinted windows pulls into the parking strip and sits in idle; maybe a chef, maybe a junkie. I stare out at the view for a moment, the gray clouds sitting low enough to hide any changes made to the skyline in the intervening years. But things have changed. Now, the opioid crisis is everywhere. Unconfined to back alleys and city parks, it fills the streets with homelessness, the hospitals with overdoses, and quietly dozes in middle-class living rooms. I squint into the gray, trying to envision a future under sunnier skies, a paradise unpaved, where public art is brightly colored, children play undisturbed in the long grass, and food flows downhill like water from a community garden atop Council Crest to the people below. But the vision fades and I turn and walk away from the present tableau; leaving all of them: the family at its picnic, the crippled mother and her flying child, the junkie and the chef, behind me on the hill. Perhaps in another twenty years I will finally get a different view from Council Crest.

WHAT LIES BENEATH

B Y THE TIME I WAS SEVENTEEN YEARS OLD I COULD EASILY be described as an environmentalist; more than that really, a full-fledged idealist, a true tree-hugging dirt-worshiper. I believed in a better world, the preservation of natural resources, biodiversity, sustainability, and the possibility of a healthy planet. I believed that grassroots organization, social change, thoughtful public policy, and science and technology could and would conspire to protect the natural world from the consequences of modern society and reverse, at least in some part, the damage we had done.

A child of the eighties, I received the first wave of mainstreamed environmental education, learning early about the hole in the ozone layer, the need for recycling, the devastation of the rain forests, and the far from accepted notion of the greenhouse effect. By the time I turned eleven in 1989, Gallup polls showed that seventy-six percent of Americans considered themselves environmentalists and that the environment ranked as the number-one voting issue. A daughter of Oregon, I was sensitive to issues of forestry and sustainable practices. And I, like many of my cohort, had been indoctrinated into the regional gestalt, a kind of Ecotopian exceptionalism, which held that the Oregonian quality of life was deeply tied to our abundance of natural beauty and resources. As a rule, we thought little of the fact that our beaches were public, gasoline unleaded, bikeways plentiful, and bottles recycled, ignorant to the radical and unique nature of these recently-won legislative battles. We had been taught, and believed, that environmentalism was about restoration, that somehow, the environment was repairable.

It would take more than ten years for me to come to question this paradigm. Ten years to understand that environmentalism cannot be about restoration, that the window for reversal has closed, that now, it is a matter of survival. It is a radical attitude in the environmental movement, and an unpopular one. Hope, it is thought, is the greatest motivator. Everything can be saved.

How does a young environmentalist come to this conclusion? For me, it played out in water, as a story in four parts about the loss of hope and the recognition of consequences. It is a cautionary tale to those who would believe that all things can be restored, a call to action with a simple slogan: Before It's Too Late.

PRELUDE: SIMPLE FACTS

AS AN UNDERGRADUATE GEOLOGY AND ENVIRONMENTAL-SCI-ence major, I took a lot of gloom-and-doom classes. I sat through entire courses on dwindling biodiversity, the plundering of natural resources, and the plight of the polar bear. I got used to the ideas of extinction, acid rain, and overdevelopment. I began to compartmentalize the problem as my professors did:

Here is environmental injustice, here is deforestation, here are dragnets. The process of fixing the environment became just as fragmented as its destruction.

Except for one thing. Every class, eventually, pointed out the following:

- Less than three percent of all the water on the Earth is fresh.
- More than fifty percent of our bodies are made up of water.
- There is no living organism nor natural system that does not, in some way, rely on water.

And then one day, in an introductory hydrology class, a single slide showing a trickle of brown water coming to a stop in what looks like a desert mudflat. In truth, it is the parched base of a dried-out riverbed. The caption read: "Termination of the Colorado River north of the U.S./Mexico border. The river, which once discharged into the Gulf of California, now runs dry before ever reaching Mexico. This is the direct result of human use." I look again at the photograph, taking in the scene, the vast stretch of flat brown earth yawning into the distance, the ocean not just out of reach but out of sight.

With that, I had a singular and embarrassingly obvious realization. The water that comes out of the faucet? It comes directly from the rivers. The rivers? They are running dry. We are really running out of water, and not at some distant point in the future. We are running out of water now. Still, there always seemed to be more pressing issues, logging protests, pesticides, GMOs. The water thing, it just wasn't exciting. Besides, clean, drinkable water still comes out of the faucets and the rivers are still invitingly swimmable, how bad could it be?

ACT 1: HIDING IN PLAIN SIGHT

FAST-FORWARD TO GRADUATE SCHOOL. IT IS A SUNNY AND CLEAR fall day in the La Plata Mountains of southwestern Colorado. *La plata* means silver in Spanish, but the name is not in reference to the silvery-gray slopes or snowcapped peaks of this Rocky Mountain sub-range, the traditional homeland of the Ute people and the northernmost extent of Navajo territory. The La Platas are part of Colorado's "mining belt," a broad swath of thousands of mines, mostly abandoned now, that follow a southwest-northeast trending gold and silver-rich ore deposit.

We, a ragtag bunch of graduate students on a three-day field trip to collect surface water samples for our aqueous geochemistry class are ignorant of this. We believe we are in some of the most beautiful and iconic country the American West has to offer. The river canyon looks down toward gently sloped meadows, while above us under a turquoise-blue sky tilt craggy peaks ringed with scree piles and sturdy blue spruce. Dense oak brush line the banks of the Mancos, which gurgles knee-deep over fat, round cobbles. The sun shines crisply through the thin air. Places like this are why we went into geosciences in the first place, a chance to experience what we think is the untouched wild.

At the edge of the stream, we are set to the simplest of field tasks, the collection of clean samples and measurement of temperature, and pH, how acidic or basic a liquid is. An hour later, however, we are still standing, shivering, ankle deep in snow, listening to an hour of protocol demonstrations: how to filter samples and use the meter, labeling nomenclature, and field wet chemistry analysis procedures. We are just starting to break apart into work groups when our professor stops us. "One more thing guys, try not to actually touch the water, and don't get it on your pants." Sure thing, professor, whatever you say.

It is hard to not touch the water. The Mancos is swollen with runoff produced from snowmelt not far upstream from our location. Predictably, sediment and bacterial counts are low. It is the kind of mountain water you wouldn't think twice about drinking. Except according to our tests, the water in the stream has approximately the same pH as battery acid.

The low pH of the stream was not due to acid rain, illegal dumping, or an industrial release; it was caused by acid mine drainage. When mines are dug, particularly open-pit mines, the underlying bedrock is exposed to surface water. As the water passes over the rock, it reacts and releases the sulfur that is commonly found with the metals in the rock. The sulfur ends up binding to hydrogen, effectively turning the stream to sulfuric acid. There are few observable indicators of this process.

We learned the reality of this the hard way. Waiting to glove up until after we checked stream flow, our hands quickly became burning red and breaking into a raised rash. The next morning those of us that were less than careful woke to jeans riddled with holes, the clear mountain water having eaten through our pants overnight.

That morning, I stood next to the stream fingering the new hole in my pants and thought, *I would have drunk from that stream. We all would have.* For the first time in my life, I was afraid to touch the water.

ACT 2: THE NEW NORMAL

THREE YEARS LATER I AM SITTING IN THE BACK CUBICLE OF AN environmental geology consulting company. I am an underling, newly graduated with my master's degree, and not yet trusted with field investigations or remediation projects. I am on Phase I, record reviews. Today my desk is awash in paper. I am reviewing a site on Bainbridge Island, a picturesque and beloved bastion of recreation for the wealthy just a short ferry ride west from Seattle in Puget Sound. Bainbridge one of the largest islands in the sound. Characterized by its abundance of coastal landforms, its coastline is a jumbled array of inlets, bluffs, dunes, tidal flats, and rocky outcrops formed by the action of tidal waters against its poorly compacted glacial sands and gravels. To the west rises the white ridges of the Olympic Mountains, to the east, a string of conical Cascade Range volcanic peaks, Mount St. Helens standing squat to the south, its slumped scar and broken top yawning out toward the Seattle skyline across the water.

This is my first such review. I have been told to trace the history of the property and its adjoining properties back to before their first developed use. I am also tasked with compiling a list of all the contamination sources in the vicinity and evaluating their potential impact to the subject property. The initial environmental data report generates close to a thousand contamination sites within a mile radius of the property. The Wyckoff Superfund site is one of them.

I start reading the environmental report for the former Wyckoff Wood Treatment Facility; at close to three hundred pages it is a drop in the bucket of the thousands of pages of documentation regarding groundwater and soil contamination at the site. It is the broad strokes, a brief history that focuses on the site's use as a wood-treatment facility. It says nothing of it being the traditional home of the Suquamish and Coast Salish people. Nor does it mention that its long-gone eelgrass beds provided essential marine habitat for the sound. But it does present a large body of previous investigation detailing the amount and nature of largely creosote-related toxic compounds that were released to groundwater and sediment at the site. The conclusion outlines the potential for migration of these contaminants, and the current, and costly, remediation strategy.

The approach is robust. Large-scale remediation, physical barriers, and restrictions on developed uses, pump and treat, the works. The cost, paid for with public funds, is astronomical. It appears on paper, after two hours of review, that every possible attempt is being made to clean up the site. The last page of the report contains a single summary conclusion, which I can only paraphrase here as: "It is our recommendation, based on the limitations of current remediation techniques and the extent of contamination at the site, that the current remedial measures be implemented at the site in perpetuity."

The reality of this was overwhelming; even with our best technology, even with continued implementation and funding and even foreseeable improvements and innovations, there is no cleaning up this particular corner of the world. Worse, though, was the conclusion I was meant to draw: that the contamination to our property from the site was considered not a risk but a baseline condition, posing no more risk to our site than to any other site in the vicinity; that all that could be done was being done; and that our site, like many others, would slowly absorb the impacts over time. It is one of many such Superfund sites located on Puget Sound.

I realized then that this kind of large-scale industrial pollution was the new normal. The goal was no longer remediation but the management of risk. The waters of Puget Sound cannot be restored to their untouched quality. For Bainbridge Island, like so many other sites, it was already too late, there was no going back.

ACT 3: CLOSER THAN YOU THINK

IT'S A FEW YEARS DOWN THE LINE. I AM STILL CONSULTING, BUT have been moved out of the cubicle and into the field. It is a far cry from the mountains of Colorado. Now, the field usually consists of abandoned industrial sites, parking lots of dry cleaners, or, like today, the playground of an elementary school in a rural town in Washington State. Now, I have my own drill crew, my own field protocols, and at any given time a dozen or more sites in various phases of investigation and cleanup. It is repetitive work that progresses in near-geologic time. Each site requires months of sampling, report writing, meetings with regulators, and finagling of property owners, liable parties, and lawyers before even considering implementing a cleanup regimen. The idea of remediation in perpetuity no longer seems alarming, more like a reasonable assessment. That the water beneath us is not what it appears has finally sunk in. I approach each project warily. Test first. Assume the worst. I believe that I am not only open-eyed but hardened, reconciled to the enormity of the damage we have done to our planet's connective tissue.

It is 6:30 in the morning, three hours before the students at the elementary school behind which we are set up will emerge from their classrooms for recess. Three hours before they come tumbling toward us, bright balls of energy, wrapped in raincoats and sent out into the schoolyard to jump in puddles. I am here for Phase II work, preliminary soil and groundwater sampling to determine what, if any, contamination exists, and if so, to what extent. We are hoping to be gone by then, in and out without raising concern. The Phase I for the site hadn't rung any alarm bells. This was considered a formality.

It's common for public buildings, especially schools, to be under scrutiny, and common for them to have contamination as well, usually related to aging underground storage tanks used for heating oil or, as in this rural location, fuel for the district school buses. It doesn't typically pose much of a threat to public health, but school-age children can be especially susceptible to toxins, and a cancer plume at a school is devastating to a community, it's better to be safe than sorry. I let my mind wander waiting for the drillers to punch in the first hole, thinking how great it would be to go to school in such a beautiful place.

"You think we've got something here?" I hear my driller say, which was odd, since drillers are known for reliably alternating between sullen silence and a steady stream of expletives, and the good ones, which these were, knew not to influence the sampler. Their job was to produce undisturbed samples, perfect snapshots of the soil in cross section, without contaminating them on the way to the geologist. My job was to evaluate the soil as it came out of the ground, carefully choosing which layers to sample based on my field tests. My field methods have, by now, evolved. My station is covered with field meters that can tell me everything from dissolved oxygen to the presence of volatile organic compounds in seconds. I know that most of the time, there is some level of contamination. I wouldn't be there if there wasn't. I never forget to glove up. Some days, I wear a respirator just to sample soil, but today I didn't see the need.

When I turn to look at my driller, he's standing by the sampling table holding the casing carefully level with one hand, the auger in the other. Heavy black oil is pouring out of both ends of the sampler. Not soil. Not water. Just thick black oil with a carcinogenic odor so strong I can smell it from more than ten feet away. My eyes flick to the windows of the school; the lights are on now; the kids are arriving. By recess, I know, I will be dressed in a white hazmat suit, covered nearly completely in oil, and will have cordoned off the lower yard with caution tape.

I feel the last vestiges of the restoration paradigm shift beneath me, and a single dark question forces itself into the forefront of my mind. "How many of those kids have cancer?" I ask the driller. "I don't know," he replies, "but by the looks of it, we're way too late."

WASTELANDS

A LOT IN LIFE COMES DOWN TO THE STORY THAT WE TELL ourselves. So much of how we are in the world and the choices we make are rooted in the things we imagine to be true, our first impressions, superficial perceptions and assumptions. Most of the time the story is wrong.

Geology is a lot like life.

Geology is always a science of imagination, but the study of geology, particularly in the early years of one's education, can be a faith-based experience. It requires one to picture forces like heat and pressure greater than our bodies and technology are able to withstand, often happening over time periods that exceed our life spans, all of humanity, or even the current configuration of the continents. The study of the earth's materials, how they are laid down and subsequently disrupted, deformed, and refashioned almost always occurs at scales either too large or too small to be grasped in their entirety by the human psyche. Much of what you study as a geologist is non-surficial. It lies beneath the surface, at some great depth far beyond the reach of excavators or drills, and what is exposed at the surface is often obscured by surface materials and human development.

A lot of geology is about taking what is seen on the surface and telling yourself a story about it.

In the fertile far west, nearly everything is covered with dense forest and deep layers of soil and sediment that obscure direct observation of even the most surficial rocks and processes, limiting one's understanding to textbook pictures and what little can be seen from cliff walls and road cuts. This obfuscation creates a gap in understanding that can only be filled with fieldwork and experience, the essential components of any geologic education. It is a tenant of the science so important that it spurred the common adage, "He who sees the most rocks wins."

To be considered a true geologist then, one must see rocks. For that, most university programs have a simple solution—field camp: a two-to-ten-week intensive field course surveying as many of the rocks and field techniques your program could cram in. It is designed to be the culmination of one's studies.

At the University of Oregon, field camp meant you crammed eighteen of your cohort, several graduate teaching fellows, and a rotating roster of professors into vans and drove east, into the desert and grasslands of eastern Oregon, Idaho, and Montana. East, out of the forest and into the frying pan—heat, dry thunder, and rattlesnakes of mid-summer in big sky country.

The company of these relative strangers and close quarters of the van grow old nearly as quickly as the radio stations fade out, some time on the second day of driving. And just when you've decided that you have arrived in the true middle of nowhere, the van stops and you get out at someplace called Craters of the Moon.

Craters of the Moon is a seven-hundred-square-mile field of lava flows that lie along the Great Rift of Idaho, one of the best examples of crustal rifting in the world and the deepest rift opening known on Earth. Almost every type of basaltic lava feature is represented there, laid out in stark black along fifty-two miles of broken ground, land that has been stretching and thinning over 30 million years, occasionally becoming so thin that hot magma from the Earth's still-molten interior begins to boil and breaks through to the surface. It's a reasonable place to bring a group of budding geologists.

To the casual observer though, it's really just an oven-hot field of messy black rock punctuating what is sure to have been a long drive, regardless of point of origin or destination. No one stops there for the rocks. They stop at Craters of the Moon because of the astronauts.

Americans, especially those of the Boomer generation, love the space program. More specifically, they love the space program of the 1950s and '60s, the Cold War space race rich in drama and heroism. The race that was triggered by President Kennedy's September 12, 1962 speech calling for Americans to land a man on the moon and return him safely to Earth before the end of the decade. They resonated with his now-infamous rally cry, "We choose to go to the moon, we choose to go to the moon and do the other things, not because they are easy but because they are hard."

Kennedy did this at a time when space was all imagination, a vast territory of unknowns, and the reason he gave for pressing into it was simple, progress, the betterment of mankind. Or as he put it, "We set sail on this new sea because there is new knowledge to be gained, and new rights to be won, and they must be won and used for the progress of all people."

It worked. Kennedy had tapped into the mid-century mindset that believed in the joy that could be derived from technology, better living through chemistry, and the promise of self-cleaning houses and flying cars. The Boomer generation looked at the space race as a source of salvation and saw the promise of innovation. They imagined faster travel, instant food, freedom from chores and computers that would take us not just to the moon but someday navigate our way to the local mall, turn on our lights upon request, and tell us when the milk in our refrigerators had gone bad.

But first, we had to get there. And because so very little was known of space, or even of the surface of the moon, NASA scientists had to imagine what the conditions might be. And so, to prepare the Apollo astronauts for the moon, to ground them in some kind of reality, to give them a terrestrial corollary for their experience, they took them to Craters of the Moon. There, NASA hoped the astronauts, Alan Shepard, Edgar Mitchell, Joe Engle, and Eugene Cernan, all pilots, not geologists, would learn to look for and collect valuable rock specimens in spite of the unfamiliar and harsh environment.

When Neil Armstrong and Buzz Aldrin set foot on the moon in the summer of 1969, it was immediately clear that Craters of the Moon was nothing like the actual surface of the moon with its silvery dust and pockmarks. While the craters found in Idaho are volcanic in origin, the real craters of the moon are caused by meteorite impacts, the same space debris that rains upon our roof but from which our atmosphere protects us.

It didn't take long after that first moon landing for Americans to realize that what there was on the moon was not a whole lot. It was filled with stardust and lava rock, things of little interest to anyone other than geologists. And by 1970, when the first of my generation were just being born, Americans were increasingly turning their attention to more pressing terrestrial matters. The environmentalism, back to the land, alternative technology, and deep ecology movements were all on the rise and none of them were interested in the moon. In fact, Dennis Hayes from Camas, Washington, not so far from Craters of the Moon, who organized the first Earth Day in 1970, is quoted as saying, "We squander our resources on moondust when people live in wretched housing."

But when the shuttle program began in 1981, nostalgia inspired our parents to instill in us the same sense of awe and patriotism they had felt decades previous when space had been a source of primary wonder, an expansion of consciousness almost as potent as their acid trips and far-out music. To this end, they bred us on *Star Wars*, *ET*, and the new *Jetsons*, casting space for us as a place of flying bicycles, grandparents in cosmic cocoons, robotic maids, and pint-sized wise men.

And so, as children of the eighties, we turned our faces to the sky, but like Craters of the Moon, what we saw was nothing like our parents' generation had imagined. By 1984, as we watched the Challenger explode from our classrooms, we were already used to seeing color images of Jupiter and Venus, and even Earth, our own tiny blue dot in space. The hole in the ozone layer was discovered in 1981, just two years after Ronald Reagan's announcement of the Strategic Defense Initiative, a defense to the newfound capability of humans to affect our own bombardment from space.

We knew that in truth, the sky was a place of atmospheric holes through which missiles might fall but only teachers did. Space for us was not a last frontier, a technology race, or a brave choice in difficult times, but just another battle line. It was a place that had always been inhabited by people, a source of commercial and technological innovation, the landscape in which humanity would orchestrate its own demise. And the moon? It was nothing but a wasteland.

Standing at Craters of the Moon two months before the towers would collapse on 9/11, with the flat of Idaho expanding in all directions and the midsummer heat bearing down on us, a wasteland was really all we saw. It wouldn't be our last.

Four weeks later, on our return trip, we tumble out of the van, thinner, dirtier, even less enamored with our now-familiar companions but having seen considerably more rocks. We gather around our professor waiting to see the map of our location, about which all we know is that we will be dry camping, and we will be here for weeks. But the image in front of us is not a map. It is a satellite image of the United States at night. In it, we can see every touch of humanity on the land, strings of cities along the coasts, clusters and networks of lights wending their way through the interior following what we knew were roads and waterways. We were located, she informed us, toward the center of the largest dark spot on the image. "If there is a middle of nowhere," she said, "welcome to it."

That place was Christmas Valley, Oregon, which, despite its hopeful name, is a place that no one loves. There is no water, no services, no development of any kind aside from roads that do nothing more than pass through it. There is no topography, no sweeping vistas, no opportunity for recreation, not even a trail through the rugged lava field. Even the rocks themselves, in contrast to Craters of the Moon, offer nothing of note, no shining pahoehoe ropes, no glistening minerals or tree casts, just plain, black rock of "nothing-to-comment-on-here" texture and wide-open sky filled with dry desert air reaching into the horizon.

It is the definition of a wasteland, having nothing to offer, without beauty or the possibility of redemption. We were there to map its two defining features—Hole in the Ground and Crack in the Ground. It would prove to be a difficult two weeks.

First, Christmas Valley is an easy place to get lost, the vastness of absence reaches out in all directions, leaving few landmarks with which to orient oneself. The convolute nature of the terrain is shocking too, as deep fissures and folds of the lava even out across a distance, tricking you into thinking it is level. In practice though, crossing the lava plain is more about scrambling and climbing, rising every twenty or so feet atop another folded crest only to discover your facing has shifted with the aspect of the rocks, forcing you to take another bearing, but off what? There is nothing in sight. By noon, the already sharp rocks become searing hot; each day's hike erasing some of the rubber from your soles.

In the afternoon a steady hot wind combined with the sun a radiant heat from the rocks to create mirages in the distance. We spent those hours mapping from aerial photographs, sitting atop the rises of small red cinder cones that dotted the field taking turns napping, hats pulled low over faces, leaned back against a large rock in a way we had rapidly grown accustomed to, long brown legs crossed at the ankles, or simply throwing stones.

Occasionally, a lizard.

But nothing more. I have no stories of my time there to share with you. Just long days of heat and rock, blending into one another, the sun marking time across the sky, even the shadows short.

Not much has changed in Christmas Valley since that summer. There has been no push for development, no trail or road system installed, no crush of selfie-taking tourists. No one is arguing over it. It remains, in matter of fact, some of the least-touched wildland in the United States. But it is a preservation by means of abandonment, not accolade or special interest as the environmentalism of the late twentieth century would have us believe is necessary to justify the setting aside of a parcel of land from the touch of humanity.

The same thing could have been said of us, idly throwing rocks in Christmas Valley with nothing of any real value to offer at that age and station in our lives other than an openness, a promise of potential, of youth. How were we to know in those long hours of heat and rock, on the brink of first 9/11 and then the Great Recession that we ourselves would become a waste-land, a generation destined to fall to phones and opioids, whose children would see the stars only through screens?

So much of what was promised us has come true. We talk to our houses, have robots that sweep our floors, and almost everything we own, from our cars to the milk in our refrigerators, is mass produced, packed with technology, and pumped full of chemicals. And it is killing us. The Boomers, for all their better living through chemistry have become the first generation to have shorter life spans than their parents, a trend that is anticipated to continue. And now, to the generation coming behind us, it has become clear that the whole of Earth is in danger of becoming a Christmas Valley or a Craters of the Moon. Dried up. Barren. Ultimately devoid of life. Ironically, as these fears grow, we look increasingly to the sky, to the moon and Mars, in the hopes of colonizing a new frontier in a new, untouched place, a paradise of our own design. Fittingly, this new Space Race is not between nation-states but between corporations and the fancy flights of a few ambitious billionaires—men that are able to spend hundreds of millions of dollars earned through conspicuous consumption on the notion of salvation in a built-environment in space. This, when one of every thirty American children are homeless.

Are we really ready to discard our own future as having nothing of value and no potential as we have Christmas Valley and Craters of the Moon? Shall we call every place a wasteland and abandon them as well? As we do our landfills, nuclear test sites, and toxic pools? As we are doing to Flint, Michigan? Is this ending inevitable?

I believe there is another story we can tell ourselves. To begin with, Craters of the Moon, Christmas Valley, and all of Earth really are actually quite similar to the moon after all. Those samples collected during the Apollo missions show that the moon and Earth have remarkably similar geochemical compositions. So similar, that researchers believe the moon was formed when a cosmic body the size of Mars impacted the still-molten Earth, sending a glob of material out into our orbit to slowly cool into rock. So really, the moon is just an extension of the Earth, sharing our galactic positioning, abiding the same expanse of space and time, and enduring the same bombardments and environmental extremes as our blue dot, only without any of the resources or protections we enjoy. Both celestial bodies, collecting stardust.

What if, instead of assigning value to a place based on beauty or even utility, we did so on its ability to endure, its capacity to abide, to exist through time, to wait, stand sentinel, and retain its dignity; to erode and age and crease and weather and be weathered? What if the true value of a place is shown in its ability to erode and age and crease and weather and be weathered? What places then become our sacred grounds? The dumping ground whose ponds still support a row of ducklings through the water? The clear-cut? The ghost gardens of abandoned cities? The land that holds a skyscraper into the air? That remains when it has fallen? The barren rock that is nothing but the promise of soil?

One way we measure the value of our own lives is by our ability to endure. We are all striving to abide, to withstand the universe's bombardment without yielding. And that, at the very least, is what our terrestrial landscapes can teach us. For surely the goal is to continue.

So let's change the story. Instead of manufacturing a new biosphere in a far distant place, let's look for redemption, regrowth, rebirth, and renewed vitality on our own, broken planet, knowing that it will transpire on timescales and due to forces much larger than ourselves. Because if you look closely, you see that Craters of the Moon and Christmas Valley, they aren't wastelands at all. Neither are they barren or lifeless. They are filled with pockets of nourishing soil, enclaves of coolness and water, and a great diversity of intrepid plants and animals that have adapted, slowly, over time, to these foreboding terrains. And once you see that, you see that in our wastelands is where hope resides.

Hope, the stuff of stardust and soil, the promise of life, our own continuity.

HOUSEKEEPING THE COLUMBIA RIVER GORGE

THE COLUMBIA RIVER GORGE IS AN EIGHTY-MILE-LONG river-carved canyon that cuts through the 17-million-year-old basalts of the Columbia River Plateau. The plateau was formed by a series of lava flows originating from a volcanic source near the Washington-Idaho border. The basalts, like the river, cover massive portions of Oregon and Washington, continuing all the way out to the Pacific Ocean where they can be seen as far south as Yaquina Head. Mapped from above, the flows look like a Rorschach test, a giant ink blot seeping out to sea, the gorge a thick scar cut across it east to west by the persistent action of wind and rain. The plateau itself is not level, but slopes gently to the south, not enough for the casual observer to notice but enough to cause the unsupported cliffs on the Washington side of the river to continually collapse in landslides. The Oregon side, buttressed by the down-slope, hosts steep cliffs and waterfalls, over seventy of them, fed by the abundant rain. It is the highest concentration of waterfalls in the continental United States.

It is a land where what is at stake is always the view and questions of who is to maintain it and who gets to enjoy it are always at the center of the debate—an argument that is played out even between the states themselves. Unlike Washington, which allows development in many portions of the gorge, Oregon has reserved most of the area as protected lands, so to live on the Washington side of the gorge is to have near-pristine views of the Oregon side of the gorge.

It is a fierce landscape in which to live. There is frequent snow in winter, intense heat in summer, and in the western reaches, rain at all times of year. From fall to spring, there is what locals call the east wind but are properly referred to as the Gorge Gap winds. Gap winds, most simply stated, are those that result from the land acting as a whistle. Cold, dense, Arctic air pushes south and flows into the gorge like water, pouring sustained winds of sixty miles per hour down the valley for months at a time, pummeling Portland and plunging its average temperatures ten degrees or more below that of more sheltered cities to the south. It is not a wind that comes in gusts, but as a faucet, turned on by the temperature difference between the eastern and western ends of the chute. The winter east wind freezes everything it touches. It's common to wake up near the mouth of the gorge to find the entire world encased in sleek clear ice more than a quarter inch thick. By mid-winter the falls on the Oregon side have all grown icy beards. For six long months, the cold gets in your bones.

I know this because in the 2000s I lived for a time on a piece of land recently acquired by a local land conservation group with the intent of converting it into a viewpoint for a popular hiking trail, the Cleveland House on Cape Horn. Today, the house is gone, in its place a wildly popular trail that culminates at a cliffside viewpoint, what used to be my living room.

To reach Cape Horn, drive Highway 14 east along the Washington side of the gorge. Just beyond Washougal the highway climbs up and away from the river and onto the plateau. From river to crest in the western portion of the gorge is about 1,500 feet of elevation gain, though the highway meanders so much that it is hard to tell that you have gone uphill at all until you turn a corner and find yourself looking out to the east across the gorge from an eagle's perspective.

The Cleveland House was down a quarter-mile gravel road flanked by fields of wildflowers and grass that led to a small rise that obscured the view. At the crest of the rise the road split, one track leading to a horse corral and stables, with a full barn and servants/laborer accommodations. The second track pitched steeply past a half-acre fruit orchard teeming with ripening plums and sad, under-groomed pear trees. Eventually, it spilled onto an asphalt-paved parking area in front of a four-car garage that blocked all but the top half of the massive blue-gray structure which sat like Humpty Dumpty, tottering on the edge of a near vertical cliff. From the front drive, the house entirely obscured the view.

The house was constructed before the Columbia River National Scenic Area was established by Ronald Reagan in 1986, protecting 292,500 acres of undeveloped land in Oregon and Washington. The early 1980s had been a period of rapid growth and development in the gorge and especially the Cape Horn area, which, with its spectacular views and relatively close proximity to Portland, was early platted. When the economy turned and the houses went up for sale, the Forest Service, in conjunction with the Public Land Trust and a few key private investors, started purchasing the land around Cape Horn for public use. In 2005 only the Cleveland House remained. Now the house stood empty, abandoned by its previous owners unable to afford it, an early warning sign of the coming housing collapse of 2008. The long-term plan was to convert the bluff from a residence to the pinnacle lookout of an eight-mile loop trail. Someone, the interested parties agreed, should keep an eye on the property until it could be demolished. For that, they found us, a scruffy trio of unattached twentysomethings that worked in wildland restoration and environmental remediation. When we moved in that summer, we knew we were there to see the house to its destruction.

The grounds were a monument to man's desire to exert control over nature and stood in direct contrast to the natural beauty of the gorge. There was a koi pond, which fed a concrete and fiberglass stream that flowed under a small arched bridge leading to the front door. The stream meandered around the side of the house, through both a wishing well and a small stone-carved fountain shaped like a giggling cherub before navigating a series of man-made terraces that comprised the backyard. Eventually, it made its way to the edge of the cliff, not more than twenty feet from the back of the house in places, and followed a footpath more than two hundred yards down the rocks to a Greco-Roman tile soaking pool that perched on the side of the cliff. From the back of the house, nothing obscured the view.

The eastern face of the house was three stories of glass, each with a massive deck. Inside, the top story was a five-hundred-square-foot master suite that was lofted, opening out into the main foyer of the house, looking directly out two stories' worth of Columbia River Gorge. The rest of the inside was overseen by a giant stuffed elk head mounted to the opposing wall.

It was akin to looking at a fun-house mirror, with everything at a funny size and angle to its real self. Whole rooms stood empty. There was an entire wing of the house we never used, three empty bedrooms, two full baths, and a library. The living room housed two couches, three coffee tables, and a foosball table and still looked cavernous and empty. There were three wood stoves, which was important because there was no way, even with three people with good jobs, that we would have been able to afford to heat it properly. In the dining room, we ran three tables together and sat, like Citizen Kane or some strange tea party, at far reaches from one another. Behind the house was a large barn with a full apartment, presumably servants' quarters of some kind, and enough stables for us to house up to five horses. It was a monstrosity of new wealth and bad taste complete with Jacuzzi tubs, walk-in showers big enough for two people to simultaneously bathe undisturbed, kitchen counters long enough to nap on, and a safe room.

We had none of the wealth or privilege to allow us to live in such a place by our own means, but live there we did, far removed from the pulse of normal city life, having little cause to leave the property for days at a time. There was no television, no internet, no local pub or coffee shop, not even particularly good phone reception. What was in abundance was land, time, and visitors. It became a kind of guesthouse, welcoming overnight visitors either too weary, drunk, or simply iced in, to head home for the evening. Inevitably, there were parties. On weekends I entertained, hosting dinners, cooking brunches for hungover guests, and leading hikes. Mostly I allowed people to stare into the distance, looking out from the developed edge, at Oregon's half of the riverscape, down the gorge towards Multnomah Falls, which you could just make out on a clear day.

Multnomah Falls is the fourth-largest waterfall in the United States, its cascade dropping 620 feet over two tiers of hexagonal basalt columns into a broad, shallow pool at its base. A trail switchbacks up the face of nearby slopes, allowing access to the upper face and crossing the water at a stone footbridge just above the second tier. The stream is derived from underground springs fed by the infiltration of rainwater around nearby Larch Mountain. In the spring and fall runoff and snowmelt contribute to the flow, engorging the falls and encouraging a revolving wildflower bloom. All year hundreds of thousands of tourists flock to its base to stand in the spray and take family photos.

It remained one of the most popular attractions in Oregon until the late 1990s when casinos began to draw more visitors than trailheads. It is also a City of Portland park, even though it is located nearly thirty minutes beyond the city limits, the only municipal park outside of city boundaries in the nation. The story of the falls and how its odd protected status came about is one of many tangled histories in the gorge, like that of the Cleveland House, where homesteading, industry, and development have been pitted against conservation and natural forces.

Often, it all hinged on a woman.

The land encompassing Multnomah Falls was originally claimed for white people by William Griswold in the mid-1800's, who homesteaded the land with an eye to logging. Gorge land, while inhospitable, offered easy access to timber, fisheries, and river transport, both by ship and rail, and was already developing with those things in mind. By the early 1900s, however, wealthy weekenders from Portland had made a habit of visiting the falls to escape the bustle and grime of the city. As Griswold pushed for increased logging and began to gather investors to fund a mill, Portlanders pushed back, blocking him at every turn. Eventually, he pulled up stakes and returned to the east coast. With his departure, Multnomah Falls passed on, to the surprise of many, to his daughter Jennie Griswold, a D.C. socialite newly returned to Oregon, bohemian, and most startlingly of all, a woman. Undaunted by the landscape, Jennie took a progressive view of her inheritance and, recognizing the value of the land in tourist dollars, not just trees, gave up the idea of the mill and started charging visitors ten cents each to hike the falls.

All was well until George Wetherby, a local entrepreneur, sought to reclaim the falls for industry and build a lumber mill to be powered by the falls. In 1905 George filed a claim to the fall's water rights and leased the surrounding land from Jennie for far more money than could be raised on hikers. The mill, likely a ploy to later attempt to purchase and log the land, never appeared, and the lease persisted for nearly ten years, Jennie welcoming hikers and visitors all the while.

In 1914 the issue came to a head with the proposed construction of the Scenic Highway, of which Multnomah Falls was to be the crowning jewel. By then, the falls had become a beloved fixture of Portland life, a welcome and accessible respite for those with the means to enjoy it, due in large part to the ongoing hospitality offered by Jennie Griswold. Portlanders wanted their falls. So, the city of Portland brokered a deal, finding a third party capable of buying out Jennie but who was also willing to deed the property directly over to the city upon purchase. Additionally, a state bill passed at the eleventh hour prohibiting the diversion of water from the falls for any purpose established the falls as a municipal park, despite its distance from the city.

Little of the story of the falls is common knowledge and the details surrounding the land transfer and associated legislation are not clear. The role of Jennie Griswold in particular has been largely relegated to a footnote in the story. No one really knows if she was on the side of the conservationists or the developers, or maybe just looking for the highest bidder. What we do know is that she took up residence at the falls until the land was turned over to the public.

That there is any history of Jennie Griswold at all is surprising. Few women's stories have made it into the history of the west. As Griswold and her contemporaries predate equal rights and suffragists and Title Nine, women in the west were only rarely land or business owners, and even more rarely politicians or policy makers. As a result, there are few memorials or roadside monuments to their contributions, which surely, as Griswold proves, they must have made. Of men, there are many. There are roads and schools and restaurants bearing the names of famous trailblazers like Sam Barlow, William Clark, and Meriwether Lewis. It goes on. Not so for women. Typically, those few landforms or markers that do bear the names of women do so only by their first names, like Marys Peak, or the cities of Veneta, Florence, and Athena. Sometimes, even less, as with the "Pioneer Woman's Grave" marker along Highway 35. In this way, women of the West are erased from history, appearing detached and family-less even in their own honorarium, leaving passers-by to wonder at the reference.

Even scarcer are descriptions of their day-to-day lives, their inner workings, or personal histories. What stories do exist are run ragged from the repetition of familiar stereotypes often crafted for the purpose of celebrating a particular and narrowly defined archetype of womanhood: the beleaguered pioneer wife, weary from the wagon trail and tied to a life of drudgery, manual labor, and isolation in the barren, open lands of the west. It is, I think, part of a larger misogyny of the west, one perpetuated by Manifest Destiny, the propaganda of western expansion, and the modern entrenchment of the patriarchy.

The gorge, in particular, has always been a terrain whose control and possession has been gate-kept by men. The first 150 years of its development are dominated by trappers, traders, explorers, and sea captains, men intent on the commodification and taming of what never proved to be the Northwest Passage. Even the rocks have traditionally been used as symbols of the gorge's masculine identity. For hundreds of years, a 25-million-year-old landslide remnant that sits jutting upward from the water has been described by viewers in a manner that notes its resemblance to a large phallus. In the modern era, the beach area adjacent to it, which was the first public nude beach in the United States, has been referred to as variations on the theme of "cock," the current iteration of which is Rooster Rock.

In spite of all the crowing, women did, in fact, make it to the gorge, just as they do now. If our current understanding of women is any indication, they arrived as far more than wives and mothers, which of course they also were. The very act of living in such a place implies they were curious, hardworking, and resourceful. Existing historical records indicate they were also landowners, speculators, proprietors, entertainers, and hostesses. Moreover, they were housekeepers, all of them, for the few intrepid travelers that made their way up the river, into the wind, to sit and take in the view.

Housekeeping in the gorge is an erosional process, although it's less about wiping the dirt away than trying not to be swept out with it. Likewise, development in the gorge has always been a fool's game, at best a long-shot gamble on slippery and unyielding terrain. Most things that are built there rarely stay for long. The valley and its falls are an absence, carved through millennia by the repetition of the water over the rocks. The river too, has had to change courses so many times that abandoned channels can still be seen looking east from the viewpoints. But building is what is done when there is money to be made, and the building of the gorge was no different, once the road was in.

That road. The road that is now the Historic Scenic Columbia River Gorge Highway, which was built in spite of the audacity of the undertaking, was the first official highway in Oregon, conceived of and constructed during a high point of expansion and optimism. It winds along the cliff edge between waterfalls overlooking the shifting scene below, and is dotted with viewpoints and elaborate vistas built by master masons from imported sandstone, which constituted the first publicly owned rest stops in the United States. Such a road and its vistas, especially in such an extravagant time, necessitated lodging. And lodging, for those with the means to vacation in such a remote part of the world, necessitated hospitality. So as the road appeared, so did guesthouses. Like the road, they were indulgent, over-the-top affairs, most of them adjacent to a waterfall. By the 1920s there were more than fifteen guesthouses hosting wild parties, "blows" in the swinging parlance of the time, between Rooster Rock and the Dalles sixty miles to the east, all of them with a housekeeper.

One of the earliest of these houses opened in 1912, before the road, on the point overlooking Rooster Rock, which is now the Portland Women's Forum lookout. At that time, the land was occupied by Alba and his wife, Madge Kay Morgan, who developed it with the intention of opening a public inn, rather than a private guesthouse. Mr. Morgan named both the land and the inn Chanticleer, in reference to the prideful Rooster in the common fairy tale who escapes Reynard the Fox only to learn that it is not he that causes the sun to rise every day. It was a modest lodge, built low to the rocks, and the salon was cozy, a grand piano filling the space between broad windows that looked out over the gorge. They were known for their creamed-chicken dinners, as it was the only thing on the menu unless salmon was in season. Nevertheless, it quickly became a popular gathering place; a year later it hosted the meeting that sealed the contract for the construction of the Scenic Highway. Shortly after that, Madge passed away from unrecorded causes. Two years later, Alba remarried and Marie Morgan became the keeper of the Chanticleer Inn, where she stayed until 1930, when it was lost to fire.

Shortly after Marie's arrival, Margaret Henderson, the interim housekeeper of Chanticleer left to open the Latourell Falls Chalet, but it too burned to the ground in its first year. Undeterred, she opened Crown Point Chalet the following year, up-slope of where the Vista House, what is now the gorge's most famous public lookout, is now located. Charlie Chaplin, the Vanderbilts, Henry Ford, and other luminaries of the time visited the Chalet, gazing off towards Cape Horn, just upstream of it on the Washington side. In pictures from the past it is always this way; people are not shown in their revelry. In fact, images of these parties are nearly impossible to find. Instead, there are only pictures upon pictures of people gazing out into the gorge, the view enough to satisfy them. Ultimately, there were troubles with Crown Point Chalet too, the Gap Winds taking off its roof in its first winter to name just one. But Margaret (and her house) persisted, serving chicken dinners for crowds of up to 150 until the realities of the Depression and failing health forced its closure and eventual demolition in the 1950s.

The Latourell Falls Chalet was not the first guesthouse in that location. Before Margaret Henderson, in the 1860s, the land was owned by Joseph Latourell. He was a French-American orphan who was shanghaied in San Francisco and placed on a northbound ship to whaling country. He escaped in Fort Vancouver, Washington, and fled upriver to the present town of Washougal. He eventually made his home across the river at the base of Latourell Falls, married a local woman, Grace, and opened a general store. He and Grace offered simple lodgings to travelers in a log house. They were known for their entertaining, with lively music and good food and their house became the center of a thriving community, but it burned down and was replaced by Maffetts Villa. Grace and Joseph continued to live near the falls together though for nearly sixty years, having eight children along the way. Today, their original home is the site of Guy Talbot State Park, tucked away into a side canyon at the base of Latourell Falls.

Guy Talbot was the first president of the Oregon Electric Railway, the line which connected Portland and Eugene in the early part of the twentieth century, and the president of the Pacific Power and Light Company. The Talbot family used the Latourell Falls area as a summer residence until the 1920s, when Guy donated over two hundred acres of the land to the state. While there is no historic mention of any note to his wife, surely there must have been a housekeeper. There must have been someone, some woman, to clear away the dust, make beds, cook meals, wash windows, and generally sustain through the drudgery and toil of everyday life in the gorge. Houses do not keep themselves.

A hundred years later, I was duplicating the efforts of these women, desperately keeping house as a caretaker of yet another doomed house in the gorge. I imagine that like for me the demands of the house were more than they could sustain. Houses like that are doomed failures, geologic or otherwise, even when not slated for demolition. The massive south-facing garden and the overburdened plum orchards made for weeks of harvesting, storing, and processing. The view and ample bedrooms resulted in an overabundance of visitors all requiring hospitality, entertainment, and elaborate meals. It took hours to vacuum the house, even with an entire wing closed down. But I did, lifting out pieces of the gorge blown in through unseen cracks and crevices. It was a losing battle that somehow, I was unable to let go of, the tending and keeping of the house and its view, even though I knew the eventual purpose was to tear it down.

For me, it was a time of introspection and intense loneliness intensified by visitors taking their leave. Little did I know that as I sat with cup after cup of tea looking across at each viewpoint, another housekeeper had sat and looked out across the gorge at me through a hundred years of history.

The role of women is important in the history of this place, not just as hostesses, but also as stewards. Each of these way-sides has passed, in their care, from private to public ownership as each landowner discovered the difficulties of prospering in such inaccessible and tenuous country. The gorge, these women knew, refuses to be turned to profit and demands to be experienced on its terms, not ours. Their great houses were, like the rocks, broken and beaten: knocked down, burned down, blown down, and, finally turned to the public trust, not timber.

Sadly, the stories of these women and the work they did towards crafting the Northwest culture and preserving its natural legacy have nearly been washed away with their houses, just like everything else in the gorge. Little remains of their lodges, dance halls, and sitting rooms. In their place are monuments to the men that built the roads, erected the parking areas, and signed the property transfers.

Like most of the great houses of the Columbia River Gorge, and the Scenic Highway, the Cleveland House took less than twenty years to become obsolete. We spent one winter there, two summers, one full turning of the calendar. Long enough to see the spring bloom, harvest the last of the apples and plums from the orchard, and witness the great shuddering of the house upon its foundation in the winter winds. Two years later, a friend forwarded me a video of the house being demolished—the final piece of the Forest Service restoration plan.

A viewpoint was to be constructed. We hiked it together one last time shortly after that. The wild bloom was in full force with purple lupine, pink salmon berry blooms, and foxglove all emerging at once. While I have an abundance of pictures of the view, I have no pictures of the house itself, nothing that shows the scope or scale of the estate. I kept, in fact, nothing of the house, save for a single blue tile, pulled from the demolition debris just weeks before the bulldozers came and cleared even that away.

I wonder if my work, like that of the other housekeepers, was washed away with the house, or if I will become an anecdote, a story told to children by parents standing at the Cape Horn viewpoint about the woman who baked plum pies from the orchard's fruit and threw parties in the barn.

In the summer of 2017, just weeks after the total solar eclipse, a group of teenagers set off fireworks above a waterfall along a busy trail in the eastern portion of the gorge, on the Oregon side just above Bonneville Dam, with its enormous counting window. In mere hours, it trapped more than 140 hikers in the steep-sided canyon below before fanning out across the land, engulfing beloved trails, old-growth, hundred-year-old cabins, and nearly burning the lodge by Multnomah Falls.

Oregon watched and mourned, knowing that these places were irretrievable to generations, those that had protected them and loved them, and several that would have no memory of what they once had been.

NO MEN, NO GUN

THE ALASKA TRIP WAS AN UNLIKELY ADVENTURE, THE result of happenstance, the unplanned convergence of three friends' vacation schedules. None of us had been to Alaska previously and we all lacked the funds for such a trip, but we had all managed to pony off something; a fiancé's attendance at a wildlife conference, an offer to stay at a rarely seen uncle's cabin, an unexpected companion airfare. Serendipity. The guys took the coincidence of our arrivals to the Kenai Peninsula within days of one another as kismet and planned a reunion.

We met in Seward, a port city situated at the mouth of a fjord on the Gulf of Alaska. It is known as the gateway to Kenai Fjords National Park and as the beginning of the Iditarod. It is, by any measure, a tourist town. A steady stream of cruise ships swells its population of approximately three thousand to nearly three times that daily, filling docks, streets, and whale-watching boats with droves of picture-taking tourists. We went sightseeing for a day, taking in the receding tidewater glaciers, horned puffins, and orcas before gorging ourselves on platters of fresh-caught halibut and salmon. But the guys were hungry for more than fish; they wanted an authentic Alaskan experience. To that end they had planned an ambitious forty-mile backpacking trip in the Kenai Mountains. Five days on foot in the heart of grizzly bear country. I was to be included, even though my boyfriend had sworn early on in our relationship that he would never take a girlfriend backpacking because they "couldn't handle it." As a geologist, I managed to carry enough outdoor credentials for him to make an exception, but solely as a ride-a-long. The planning and packing, he insisted, would be left to him.

From Seward we drove north out of the bustle and crowds until the houses thinned and then petered out, and into the Chugach National Forest until we had the birch- and spruce-lined road to ourselves. It was early September, just after the salmon run and near the end of the berry season. The time of year when animals hurry to fatten up before the long winter, a time of year considered good for hunting and bad for encountering bear. On the Kenai Peninsula, black and brown bear (grizzlies) are abundant. While no one knows for sure how many brown bear make their home on the Kenai Peninsula (the Department of Fish and Wildlife hasn't conducted a formal survey since 1980) it is estimated that their population exceeds two thousand. We knew that an encounter was likely.

There are two things Alaskans talk about before they know you well enough to discuss fishing and hunting: how much land they have and brown bear. They especially like to talk about brown bear to tourists, a behavior they claim is a form of public service but is really a kind of sport. In just over a week, I had already grown accustomed to stories of sturdy locals chasing brown bear off their porches with brooms or holding their ground as a six-hundred-pound blonde with a twenty-five-inch skull charged their vehicle, snowmobile, or entire family. But even with their knack for exaggeration, it was clear that Alaskans have a healthy fear of brown bear. It was also pretty clear that this fear was justified. Brown bear stand up to ten feet tall, run at thirty miles per hour, and are fiercely protective of their young. If you run into one, they will stand up and holler at you if you are lucky, and fully charge you if you're not the former. I had witnessed this behavior several times in bear country in the lower forty-eight. If you have something that interests them, they won't wait for an invitation; they will enter cabins and break car windows and generally tear things apart for something even as simple as hummingbird nectar. And their senses of perception are keen: They can see, smell, and hear you coming a mile away.

So what Alaskans do is carry guns. Lots and lots of guns. Shotguns and rifles if they are formally outside, a simple side-arm if they are in a generally developed place, which they claim will only aggravate an approaching bear. From my experience, most Alaskans do not go into the wild, or even sometimes their own backyards, without a shotgun.

I do not carry a gun. I've never even touched one, though I've had plenty of opportunities. I was raised by Catholic pacifists, have been a vegetarian—even vegan at times—for most of my life on moral grounds, and I am generally disinclined toward weapons and averse to violence, even in media. But I am also a strong supporter of the constitution and its many protections. It's not that I have an issue with guns, per se; I have an issue with how many people choose to behave with their guns. Alaskans, I have found, are generally not those people. Many Alaskans are hunters who live near and among some really big and scary animals, including wolves, bison, and black bear. Even moose, in spite of their cartoon appeal, are not to be trifled with; they stand up to eight feet tall, have antlers that can be six feet wide, run at speeds exceeding thirty miles per hour, and are pretty mean as well. In light of all this, the general consensus of Alaskans seems to be better safe than sorry. From everything I'd heard about Alaskan wildlife, I agreed.

On previous excursions to bear country I had always made-do with bear spray, best practices, and a lot of noise and so had the guys. But I was not the planner or packer for this trip, and this time we had brought a gun. It was a loaner from our hosts, a rifle of some kind that we had dubiously stowed among the rest of the gear in the back of the jeep. An addition only made after my boyfriend had dutifully gone out behind his uncle's cabin to shoot it into a dirt pile in an attempt to prove to me that he knew how to use it without taking one of us out accidentally. It was a training session that appeared to do more for his masculinity than his skills set, which I figured was fair enough. We were clearly in men's country—the kind of place where work boots are worn on weekends, the diner walls are covered in trophy kills, and everyone stands when a woman joins the table.

It was a far cry from the liberal-progressive culture from which we hailed, and the guys were in testosterone heaven. In many ways they were prototypical Northwest guys of the educated, woodsy, mandolin-playing, craft beer drinking, goatee-wearing, ultimate frisbee ilk, one leaning more towards granola, the other more towards hipster, both employed by the same conservation group. They were, by Alaskan standards, soft. Usually, they would have had nothing to do with firearms, but they had Alaska fever and were relishing the opportunity to be men's men. So on the subject of the gun, they had taken something of a bemused and intellectualized *Lord of the Flies* theme-park approach to the whole thing—somewhere between a fraternity-like testosterone binge and an anthropological study. Neither of them thought bringing it was necessary and both were vaguely bothered by their easy abandonment of their values system.

But then, neither of them wanted to carry it either, because whoever carries the gun has to face off with the bear. We had been told by locals that the trick to stopping a grizzly was to hold your ground and wait until the very last moment to shoot; get it right in the face. Anything less would just piss it off. As we drove toward the trailhead, our bellies still full of fresh-caught fish from Seward's docks, I eyed them from the back seat. They were discussing the pros and cons of bringing it with us into the backcountry, the moral repercussions of using guns for recreation, and the possible stories of heroism and bad-assery that would result from their successful standoff with a grizzly. I tried to imagine either of them holding their ground against a charging brown bear mama, and couldn't.

At a pullout along the Sterling Highway we pulled over, parking the jeep at the southern terminus of the Resurrection Pass Trail and stashing the keys in the tailpipe for the shuttle runners that would, hopefully, leave it for us at the other end. It was later in the afternoon than we had wanted to arrive, so we decided to make dinner before heading out to find camp. True to Murphy's Law, things that could possibly go wrong already were. On top of our late arrival we piled on some dense clouds rolling in and the discovery that one of our stoves was not working. It was one of those foreboding moments when you would usually start to question the wisdom of your plans. But the guys didn't seem bothered by any of this; they were still debating the gun. I listened as one argued against on moral grounds and suggested that making enough noise would alert any bear to our presence, averting an encounter. The other argued in favor, objecting to the prospect of having to hike forty miles hollering every thirty seconds. He was referring to the traditional anti-bear call, an elongated *"Aaaaaaaaaaaayyyyy beeeeeaaaaaaaarrrr"* used universally by back-country hikers to warn bear, especially mama bear with young, that humans were near. But he had an ulterior motive. Unspoken in his support of the gun was this—he was the smaller of the two and traditional wisdom holds that the largest member of the party should be the one to face off with a brown bear. This was not lost on our larger companion, who attempted to counter the hollering argument by mentioning that we also carried bear bells. In this, he forgot that they had mercilessly made fun of me when I attached them to each of our packs the day before, deriding me as a Girl Scout.

"Besides," he grumbled as a last resort, "I'm already carrying the bear spray."

"Which will only work if you let it get within ten feet of you. And by work I mean piss it off, or maybe slow it down, and that's if you're not downwind. If you are downwind the spray will just knock you out and make it easier on the bear" was the grinning reply.

"Great."

I kept quiet; my point of view on the matter was of little use. I was the smallest of the three, I had no idea how to use a gun, and even if I did, it would likely just knock me down. It wasn't going to be me lugging the thing around. I figured they would sort it out between themselves by flipping a coin. In truth, the Alaskan bear talk was getting to me and I wanted all the sense of security I could get; even if that security was nothing more than a soft Portland hipster with a rifle he barely knew how to use. Maybe our hosts were right when they pressed the rifle on us; maybe it was unconscionable to take your girlfriend into the wild without reasonable protection. What, I wondered, did all the other women do? What was reasonable protection without a man with a gun? So far on our trip I hadn't seen many women outside of towns and visitor centers, and the ones I had seen wore clearly visible sidearms—not anything that would be more than a warning shot against a bear, so it seemed clear that bears weren't the target population for that particular fashion statement. I focused on packing up the kitchen.

When I tuned back in to the conversation a few minutes later, they had decided to just leave it in the car.

And before I could hoist my pack they were headed off hollering and clapping down the trail, bear bells chiming merrily at their hips, the rifle resting beneath a blanket on the back seat of the parked car. I stood, feet cemented to the ground, watching as they disappeared around the first bend and weighed my options. The obvious choice was to sacrifice my carefully constructed competent-outdoors-woman exterior and dissolve into a hysterical and girlish fit designed to capitalize on their primal, though liberalized, protector instincts. Or, similarly, to go chasing after them, begging for them to take pity on me and bring it, like Scarlett O'Hara in the mist, all helplessness and fear. But I worried about the larger repercussions of sacrificing my dignity in either of these ways, especially if we did not, in fact, encounter any bear. By playing the girl card would I be dooming myself to the role of outdoor housewife, constantly doing dishes and filtering water while the men built fires and roasted things on sticks? Or worse—never be invited on another trip?

I hesitated. They were really making a lot of noise. They could probably be heard by every bear in the state, I reasoned. I should just buck up. What ultimately got me moving though, was not bravery or reason or agreement with them, or even fear of being left behind. No, it was the singular thing I knew to be true about escaping a charging grizzly: You don't have to be faster than the bear, just faster than your companions. Gun or no gun, I thought, I had a game plan. Off I went down the trail after them.

It was an anxious first few miles. Distances are big in Alaska; the sky is big, the mountains are big, everything is big and not in a normal way. Big in Alaska is an order of magnitude larger than in other places. The landscape leaves room for thought. As we made our way down the trail, pressing through thigh-high grasses and dense stands of Alaska blueberry and chest-high fireweed, we worried in circles and listed those worries in between bear calls and bouts of hand clapping. We discussed the guys we hired to shuttle the vehicle, if they would be able to find it, drop it at the right place, or show up at all. We tracked our miles, concerned about making good time and finding a suitable place to camp before dark. We ran over our inventory, wary of running out of supplies; we were already down a stove and low on fuel. Then one of us spotted scat on the trail and later we heard a rustling in the brush near our tents and our minds turned to bear. We went to bed before the sun set hoping to start fresh the second day, but continued rustlings got us up in the middle of the night to spend nearly an hour rehanging our food bag one hundred feet farther from camp, nearly out of sight. Bear paranoia was setting in.

The next day started out well. With bellies full of oatmeal, a few miles under our belts, no further gear failures and none of us having become bear fodder in the night, our spirits were up. We switched from bear calls and clapping to camp songs and rehashed-'80s tunes, covering ten miles by lunch, a good pace by any standard. But weather that had been threatening to move in did so in the afternoon. Heavy, dark rain clouds descended fast into the U-shaped glacial valley and fat, cold raindrops began to fall. We pulled on our rain gear and hiked fast to a hunting shelter on the map, not wanting to be soaked and cold so early in the trip, and hoping it didn't turn into an Alaskan summer snowstorm. We arrived there just as the rain began in earnest. It was already occupied by two enormous hunters posted up to watch their friend track a caribou across the opposing ridge with a pair of high-powered binoculars. These guys were not soft.

The hunters saw the guys first and called them over, asking them where they were from (Oregon), if they'd seen any game (not with all our hollering and jingling), and what they were shooting with (bear spray). Standing in the shadows cast by these barrel-chested roughnecks made my companions begin to squirm. You could practically see the machismo draining from their bodies. Finally, the hunters noticed me, and with total incredulity asked, "Wait, what is wrong with you two that you're out here with no gun *and* a girl? What are you, stupid?" I lifted my hand in greeting.

And actually, they turned out to be not-bad guys, or at least surprisingly open-minded given their cultural context. We passed the storm with them on the tiny cabin porch, listening to their hunting stories and asking them questions about the trail, the weather, and if they had seen any bear. They never managed to speak to me directly or even make eye contact with me, an interloper in every man's land, but they did eventually admit that they liked the idea of "a girl in backcountry." They just didn't think it was safe and wouldn't risk it themselves; like how other people feel about beekeeping or juggling chainsaws. One of them, in an effort to illustrate my usefulness, suggested eating me if we got caught in a summer storm, as I would surely be the first to succumb to the elements. They both congratulated the guys on being generous enough to let me tag along. Finally, as a parting gift, they promised to check the trail logs to make sure we made it out the other side.

Two days later I was hanging tough, uneaten by man or bear and the only member of the party not crippled with blisters. I woke up early, settled in to a cup of gritty coffee in the morning calm and thought back to the gun and my early doubts about the trip. We had seen bear high on a ridge above us and evidence of them was all along the trail, but we had had no close encounters with anything more threatening than the occasional marmot, an oversized tundra rodent easily scared by direct eye contact and harsh words. We had grown both less and more paranoid about the bears. As time passed uneventfully, we began to forget about them for longer periods of time, only to have the risk snap back into focus with each unidentifiable source of twig snapping sending us into bear-call mode. That morning, I was deep into my happy place of denial, enjoying the expansive solitude and play of morning light on the valley floor beneath me.

Then I heard a noise. Not a twig snapping, a really properly big noise. The unmistakable sound of something large coming at me down the trail. It took half a second for me to be on my feet, turning instinctively to outsprint the guys. And then I heard something else: a human voice yelling *"Aaaay, Bear, AAAAY, BEAR!"*

Coming around the corner were three girls. Girls! And they were on bicycles; their gear strapped to narrow trailers. Suddenly it felt like weeks since I had seen another woman. I was elated. I got so excited that I ran across the tundra, arms open, echoing their calls with my own *"Aaaay, Bear"* greeting. This scared the leader so badly she bolted upright and fell off her bike; the others scrambled to stop in time behind her, failed, and the three ended up in a tangle. I stumbled and raced to reassure her that I was not, in fact, a bear. She grabbed at her heart from the fright and we all laughed. And then we did what the guys did, asked where we were from (Montana), if we had seen any game (black bear), and what we were shooting (if pressed, dirty looks). After a few minutes they moved on and I stood watching them, jealous.

No men, no gun, no waiting to be invited, I thought as I walked back to camp. Maybe it's time for me to roast something on a stick.

CASTLES
MADE OF SAND

Everything you will ever need to know about global sea level rise you learned as a child building castles in the sand. To begin, gather a pail of dry, densely packed sand. Turn it over carefully, pressing it firmly into the ground. Remove the pail and watch it all slide away. This is the first and last heartbreaking lesson of building in sand—whatever you do, however you proceed in your construction, eventually your castle will crumble and slip into the sea. This will not stop you from trying.

IN 2008 I PASSED THE BOARD OF GEOLOGIST EXAMINATION'S test and became a Registered Geologist with the state of Oregon. It was the end of the pre-recession building boom but the Oregon coast was still exploding with development. Tillamook County, directly west of Portland, was benefitting from the frenzy of wealthy investors rushing to develop the county as an urban escape. The county, though, had strict regulations regarding hazard assessment and mitigation for new construction, requiring a site visit and summary report stamped by both a Registered Engineer and a Registered Geologist. Before my stamp had even arrived, architects and engineers with projects on the coast scrambled to fill my winter schedule. Within months I was in the coastal hazards business to the tune of ten sites a week. It was during this time that I became, perhaps not wholly, though significantly, responsible for the riprapping of Rockaway. This is the story of just one of those properties.

With sand castles, location is everything. Not all beaches are suitable for development; the nature of the sand and the shape and dynamics of the beach will play a role in the success of your project. Seek out broad, level beaches composed of well-sorted sands with angular grains; these are far superior building material as these grains will lock into place, rather than slide past each other.

IN SPRING, I WAS SENT ON SHORT NOTICE TO A RESIDENTIAL property on the south end of Rockaway Beach. For the casual observer, Rockaway is a typical blink-and-you-miss-it Oregon-coast town, little more than a cluster of rundown shops, mid-century motels, and hidden houses along Highway 101. To geologists, Rockaway is an active dune-backed beach, meaning that despite the stabilized impression imparted by development, the underlying sands are transient.

On the Oregon coast, resistant headlands tend to isolate beaches, giving each enclosed area its own sand budget. In a typical year its beaches are eroded by winter storms and rebuilt during the summer months as wave action decreases. The arrival of new sand is almost always welcomed in coastal areas. It's wasting erosion due to the action of wind and water that can prove catastrophic. In active dune-backed beaches like the one at Rockaway, it is not uncommon for their morphologies to change by tens or even hundreds of feet in response to single-storm events.

Choose a part of the shore just above the high tide line where the sand begins to soften. Map out the base of your castle with a stick and begin excavating several inches below the surface until you reach more solid, wet, compacted sand. This is your foundation.

THE *SUBJECT PROPERTY* WAS IN A QUIET NEIGHBORHOOD, A mish-mash of tiny rentals, modest, year-round residences, and, to the west, larger, view-blocking two-story vacation houses. The house was a flat-topped A-frame the same gray color as the ocean on a winter day. Like its neighbors, it was built on sand with shallow concrete footings, before zoning and building codes took long-term erosion or out-sized storm events into account, and when global warming was still far from an acknowledged phenomenon.

At the door, an older woman with a quiet demeanor and a frightened face greeted me, clasping both hands nervously around a cup of coffee. From her porch, she told me that they bought the house years before as a vacation home but had recently retired and taken up permanent residence. For all that time, she said, it had been a quarter-mile walk through shore pines, manzanitas, and grasses to reach the shore. This winter, though, the ocean had come knocking, and waves were breaking in their backyard. Now, she said, her hands shaking around her coffee, they were stuck with a piece of land that was slowly being eaten by the sea.

To underscore the point, she took me to the back of the house where they had placed wooden stakes in twenty-foot increments from the back door to what she called "the drop-off." They put them in, she told me, less than a month earlier, when the last stake out read 200 feet, but on that day the farthest remaining stake sat only 160 feet from the house. She warned me to be careful as I approached the last stake because the transition from dune to the beach was abrupt and hard to see, a swag of green-brown dune grass curling over the ledge obscured a vertical drop of fifteen feet to the beach surface. Peeking over the edge, I watched as the water from the incoming tide churned directly beneath my feet.

Water is both your castle's savior and nemesis. Too little leaves your structure a crumbling mess, too much separates the grains causing them to lose cohesion and begin to flow. Timing, though, may be the most important factor; what rests out of the reach of waves one moment can be inundated the next. To maximize your time above the water line, begin building just after the peak of high tide when the waters are receding.

IN THE PRE-GLOBAL WARMING TIME, LOCAL HIGH-WATER LINES were largely controlled by a somewhat predictable combination of wave run-up, winds, storm surges, and tides. But these are temporary, cyclical variations in water level. Global sea level rise, the semipermanent waxing of the ocean in response to an increase in the total volume of water in the Earth's interconnected ocean basins, is not. Coastal communities around the world are struggling to stay above the water.

For the last ten thousand years, since the last ice age, global temperatures have been increasing. 2011 marked the first of three times in recent years that above-freezing temperatures were recorded at the north pole, and in 2018 arctic temperatures reached more than twenty-seven degrees Fahrenheit over the winter average for nine days in a row. Two responses to this warming, the melting of ice and the expansion of water have formed a positive feedback loop. As rising temperatures melt ice from the arctic and mountain glaciers, the increased volume of water traps even more heat, while the ice loss itself causes a decrease in reflected radiation—all resulting in more water, more warming, more volume, and higher seas. Since 1870, mean sea level, measured retroactively using core samples of arctic ice, tide gauges, and remote sensing data from satellites, has increased two hundred millimeters, a rate of several millimeters a year, and recent data suggests that this rate is dramatically increasing.

The models are not encouraging. The Intergovernmental Panel on Climate Change believes that global sea level will increase by eleven to thirty-eight inches by the year 2100, impacting hundreds of millions of people living in areas prone to inundation and other related consequences. Worst-case scenarios, in which the Greenland ice sheet melts entirely (an increasingly reality-based scenario), estimate sea level changes in that time of over thirty-five feet, enough to drown all of London. The World Bank anticipates that by 2050, costs associated with sea-level-driven flooding will rise to over $50 billion annually.

The impacts of global warming on ocean waters had already begun to be obvious along the Oregon coast by 2008. In the early 2000s the Rockaway dune began to be eaten in large chunks by a series of high-intensity storms. In 2007 that process was accelerated by a series of three storms between December 1 and December 4 which would come to be named the Great Coastal Gale.

The storms were impressive, even for the Oregon coast. Wind speeds reached over 104 miles per hour for a period of more than thirty hours that saw more than a foot of rain blanketed across the region. Thousands of trees were felled, including a seven-hundred-year-old Sitka spruce which at the time was considered the oldest in the world. Forty-to-fifty-foot waves were recorded. Direct losses from the storm in infrastructure, housing, and timber alone for the entire region topped 300 million, and indirect losses were anticipated to be more than five times that. In Rockaway the sea came in and didn't go back out.

Built properly, there is little limit to the height of your castle. Pile loose sand from your excavation area over your foundation, adding water and compacting it with shovels until it reaches the desired height. Overbuilding is crucial as sand sculpting is essentially a process of removal. When your details are complete, reinforce around the base with rocks, shells, and driftwood to protect it from the fingers of water that will begin to lap at it as the tide slides in, each wave lingering a moment longer before retreat.

IN ROCKAWAY, MORE THAN A YEAR AFTER THE GREAT COASTAL Gale, it still seemed the ocean was foisting itself upon the land, each high tide encroaching farther than the last. It was an inexorable process, though that had not prevented architects, engineers, homeowners, and developers from believing that the ramifications could be avoided or curtailed—every coastal community has a long history of construction related to the taming of the sea. Often, as was the case with Rockaway, it begins with jetties.

In the late 1800s, just decades after the arrival of the first white people, Rockaway Beach was one of a klatch of short, isolated stretches of sand accessible only at low tide, collectively known as the Garibaldi Beaches. Even in those early days, savvy developers knew that if made accessible, the broad, sandy beaches, rare on Oregon's rocky coastline, would be tourist gold mines. They soon convinced investors and local agencies to construct a series of jetties that would, in theory, stabilize the beaches' sand budgets while allowing for road access. It worked. The jetties, paired with a rail line from Portland and a shore-side resort in Rockaway, made the Garibaldi Beaches a destination. By 1911 Rockaway was drawing wealthy Portlanders in droves. In the summer, housewives and children would stay for weeks at a time, joined on the weekends by weary businessmen transported by the commuter "Daddy Trains."

Remember that sand castles are prone to damage from outside forces and unexpected events: dogs chasing after balls and sticks, mean-spirited teenagers, malicious siblings, and absent-minded beachcombers.

THE BOOM TIME DIDN'T LAST, AND NEITHER DID THE BEACH stabilization afforded by the jetties. Over time, they had begun to function as resistant headlands, bending the waves on their approach and concentrating the energy between them, increasing their erosional power in the sandy bays. The Garibaldi Beaches began to erode. By 1965 the entirety of Bay Ocean, just north of Rockaway, had fallen into the sea. In 2008 the population of Rockaway was hovering just above one thousand, and even the jetties themselves were beginning to wash away. Something, people said, would have to be done.

Homeowners were on their own. In Oregon, the consequence of the ocean taking your grassy dune is the state taking your land. Since 1967, when the Oregon Beach Bill was passed, all land along the coast from the low tide mark inland, to sixteen vertical feet in elevation above it, and public easements from that point to the vegetation line have been publicly owned. It was a landmark piece of legislation nearly a hundred years in the making and unique in its time, when most beaches in the United States were held as private property.

Between 1874 and 1911 a substantive number of difficult to develop tidelands, including portions of what are now Seaside, Newport, and Rockaway were sold to the state by private landowners. By 1911 the land transfers inspired Oswald West to run for governor on a platform of public beaches, which would also serve as a state highway, thereby providing much-needed infrastructure free of cost to a state desperately in need of thru-roads. He won, and in 1913 Oregon's beaches became a highway. Tourists descended; pictures from the period show beaches crowded with revelers, Model Ts, and even the odd small airplane. The success of the highway spurred the State Parks and Recreation Department to purchase more land along the coast to build State Parks, which they did, vigorously. Today there is a State Park approximately every ten miles from the southern border to the state of Washington.

When Highway 101 was completed in 1939, the Coast Highway was no longer needed, so in 1947, the Transportation Department declared the beaches recreation areas. None of these efforts though, stopped privatization. In 1966 nearly half the coastline was still in private hands despite the state's best efforts. The solution: The Beach Bill, which, following precedent set in Texas state law, worked on the premise that because Oregonians had always believed the beaches to be public, and in fact had used them as such with state oversight, that they were, in fact, already within the public domain. The Beach Bill then, would simply honor and codify that belief.

Be watchful and keep a reserve of materials at hand for speedy repairs but be sure to pack your pail with sand from another, distant location to avoid undermining your own estate by digging too close to your foundation. Remember, do not allow other children, especially siblings, to build too close to your castle. Their choices and workmanship, the long moat your older sister digs around her land with her heel, will have inevitable consequences.

AT THE TIME OF ITS IMPLEMENTATION, THE ISSUE OF EROSION AND the subsequent loss of private land to state hands via forces beyond anyone's control, was not generally considered. After the Great Coastal Gale, all of that changed. When it became clear that beaches were changing and high-water marks were moving inland as a result of large storm events, property owners looked to the government for aid only to discover that the state had no vested interest in preserving private property lines. If the sands were to be stabilized, homeowners would be forced to bear the burden of the cost, even as their land was being converted to beach and turned over to the state. They would also be forced to pay permits and repair costs to roads, access points, and easements, which they argued, were already in the state's possession.

The cheapest and fastest solution was to harden the shoreline, a euphemistic way of referring to the practice of laying riprap, large piles of resistant rock, against the ocean-facing slope of a dune. It's a hugely expensive proposition involving geologists, engineers, heavy equipment, and specialized materials. But the real problem with riprap is, once one property owner lays rock, everyone has to. Isolated, site-specific riprap, like jetties, forms a small-scale headland, ultimately increasing overall beach erosion in those areas that remain un-engineered—passing the buck, one might say, to one's neighbors. But the storms kept coming and the dune kept failing and someone, finally, decided to lay down rock. It was in this way, eventually, that all of Rockaway was slowly riprapped.

My *Subject Property* was no different than any other in Rockaway, but something about the way the owner's hands had shook around her coffee that first day had affected me. By that time, I was hardened, well familiar with local backhoe operators and quarry managers as property owners scrambled for purchase on the dune. What did distinguish that *Subject Property* from the others was the magnitude and startling pace of the erosion, and the obvious role played by a nearby RV park, which boasted an oceanside promenade made of concrete over riprap. Whatever benefits it was providing to the park, it was bending the water towards the *Subject Property*. The wave action had carved out a sharp dogleg north of the rip rap which cut inland through the dune more than twenty feet beyond the eastern edge of the façade. It left a wall of exposed, loose sand that threatened to fail, undermining the park itself and the adjoining properties.

The beach, too, was in bad shape. To the west, pilings, driven perhaps decades earlier, now stuck awkwardly out of the sand, tipping to the sides more than twenty feet above the new ground surface. The massive drift logs that had sat on the beach during my first visit were gone, evidence of the force and depth of the water at high tide. Still, my clients, just two lots to the north of the RV park, had been unable to reach a consensus with their neighbors and the farthest stake from the *Subject Property* now read just forty feet. I would return to lay riprap, but it would not extend to meet the RV Park's promenade, leaving the adjoining houses far more vulnerable to erosion. It was a save-yourself situation.

Do not grow attached. It does no good to fight the tide, for the ocean is a nihilist. Leave the beach knowing that your castle will not make it through the night. It is the ocean only that abides.

RIPRAPPING IS NOT GENERALLY A PRECISION BUSINESS, though the process is fairly simple, especially in emergency situations, which this was. The goal: to get rock down against the sandbank. This consisted of tons of two-to-four-foot-diameter boulders being delivered by dump truck to the site where they would be unceremoniously chucked over the edge to be sorted and individually placed by a track hoe operator. For stability, each rock would need to be keyed into place by three points of contact with the surrounding rocks. It was a messy, expensive, and excruciatingly slow process that meant risking further collapse by clearing vegetation and maneuvering heavy equipment across the already unstable site.

Every time the loaded truck rolled across the *Subject Property*, I imagined the entire remaining dune failing under its weight. At low tide, we'd maneuver the CAT onto the beach from an access point several hundred yards to the north, where the dune was little more than a sinuous lump. I watched the slow progress of the CAT operator on the beach, lifting boulders with a pincher-like bucket like he was fishing for a teddy bear in an arcade lucky dip, and kept one eye always on the ocean. There was a constant evaluation of his progress versus that of the tide, a weighing of safety against the cost of mobilizing for another day. When the water came in, I knew, it would be feet deep and churning, and it would happen all at once. The CAT tore deep fissures through the sand, making such a mess that it was easy to ask if it might not be better just to let the structure go, but each day after we left the ocean would rise, scouring and leveling the beach, erasing any evidence of our tracks in its sands.

At the *Subject Property*, the homeowner was searching for meaning. On my final visit, she asked me why a thing like this happens. Everything I knew about Global Warming, sand budgets, grain-to-grain interactions, and code rushed through my mind. Putting science aside, I told her, simply, because it can. The ocean is far bigger than us, I said, and it acts on timescales much longer than human life spans or even all of human history. This shift, catastrophic to her, was imperceptible to the ocean. It was simply exercising one of its many options in response to changes in climate, season, and the shoreline. All that water has to go somewhere. An unsatisfying answer at best.

My last series of photographs from the *Subject Property* show the tide coming in, breaking against the riprap, the over-spray hitting the grassy ledge. It has been easy in the years since to imagine a full moon tide or small storm surge overtopping those temporary measures. Recent aerial images of Rockaway show riprap running continuously from the RV Park north, across the *Subject Property* and several of its neighbors. The A-frame, for now, still stands.

OUT OF
THE WOODS

N MY MID-TWENTIES I DISCOVERED A SECLUDED RETREAT IN
the Oregon Cascades, a loosely developed hot springs situated in a dense and remote stand of old growth run by sleepy
hippies—Breitenbush Hot Springs. It is a place you either love
or hate, depending on your tolerance, alternately, for hot and
cold, and the companionship or lack thereof of others. I was
hooked at my first visit.

Hot springs are natural effusions of water that emerge to the
surface at temperatures hotter than the human body. Most hot
springs are related to young volcanoes like those of the Cascade
Range. Not all the magma associated with a volcano erupts
to the surface. Much of it fills in subsurface voids and faults
where it cools over thousands of years. When it rains, water
flows down through the ground and encounters the heat from
this leftover magma, which warms it, causing it to rise back
up to the surface picking up dissolved minerals from the rock
along the way. It has long been thought that these mineral-rich
waters have healing powers.

Breitenbush was first used for recreational and health purposes by the Native American tribes of the Willamette Valley as early as the 1700s, but it was also frequented by trappers, hunters, and settlers who named it after the one-armed hunter Breitenbush, who was the original homesteader in the early 1800s. The 154 acres is held under private ownership and protected. The surrounding area though, is in a part of the largest timber-cutting district in the lower forty-eight states. Today, there is a tenuous peace agreement between the two. But for the entirety of my lifetime, they were in a near-constant battle started by what was found on a hand-built trail by a lone volunteer.

In the 1980s, stewards of Breitenbush began constructing a network of trails through the old growth with the goal of enhancing public support of their conservation efforts by providing public access to the land. It was this group that first observed a pair of nesting spotted owls. This discovery, along what is now called the Spotted Owl Trail, would lead to the eventual protection of thousands of acres of forest, but not before turning Breitenbush into a hot spot for conflict and radical direct action. That direct action, protestors blocking roads, spiking and occupying trees, and locking themselves by the neck to logging road gates, would characterize the environmental movement of the 1990s. It was a pivotal time period in contemporary environmentalism when the anti-logging movement stopped being about elusive, unprotected owls and the old-growth forests they inhabited and started being about the battle against clear-cuts. With this transition, the era of benign conservation was over.

One Christmas day I woke up early to have breakfast and open presents with my parents, then packed up my things and set off in a light snow for Breitenbush, alone. The weather made for slow going, and it was more than two hours before I turned left off the last plowed road to drive the fifteen winding miles up to the lodge.

When I arrived, the windows the of the lodge were thick with steam, and the air was filled with the smell of sulfur from the springs and dinner from the back kitchen. I was just in time for the afternoon yoga class held in the meeting hall. The room was dark and carpeted with a thick, green, all-purpose carpet. A disco ball hung idle from the center of the low ceiling, reflecting only the light from a fire that raged in the old stone fireplace, built to standing-height from native rock nearly a hundred years earlier. Outside, the snow was coming down hard and fast and I wondered how many people might join me on Christmas day, in the middle of the woods. I closed my eyes for what seemed a moment and opened them to find the room had filled with more than twenty of us.

The instructor sat and greeted us with a smile, noting that with the weather as it was anyone looking to leave in the next twenty-four to forty-eight hours was most likely out of luck. We might as well settle in.

"Take a seat," he began. "Welcome. Close your eyes. Breathe. You made it. Wherever you are coming from, for whatever reason you are here, you have arrived. There is no other place for you to be. You have a warm cabin, the forest around you, the hot springs, and we have taken care of food for you. There is nothing that you need that you do not have in this moment."

At dinner I ate with strangers, a vegetarian feast of no particular religious affiliation—previous experience perhaps informing the staff. There were many couples, some families with young children, and others, people like me, alone, in retreat, and quite content with the snow, the pools, hot tea, and the company of strangers. After eating I soaked a long time by myself in one of the upper pools, staring out at the snow as it fell. Sleep came easily in the simple cabin.

In the morning I woke up early, just before true light and slipped out in rain boots and sweats to take an early sauna. The sauna house is small and boat-like, constructed out of well-weathered wooden boards with tiny porthole windows that open from the inside. The floor is slatted and open to the stream from the springs which provides the heat and moisture for the sauna as is passes beneath the structure on its way to the Breitenbush River.

The thirty-some-odd springs that make up the Breitenbush pools emerge from the walls of the deeply incised canyon after passing over remnant magma from the Cascade's most recent volcanism. Without the action of the river, which has worked over millennia to cut into and wash away thousands of feet of overlying deposits, that heat would remain underground. But those rock layers consist mostly of volcanic tuff, a rock which, despite its resilient name, is little more than loosely welded ash and pumice. In the case of the Breitenbush tuff, it is also highly folded and fractured, making it easy to erode away. The nearness of the underlying magma is underscored by scorching spring waters—the waters at Breitenbush emerge from the ground at temperatures exceeding 180 degrees Fahrenheit.

The intensity of the heat was almost too much for me. I took deep, mineral-filled breaths and doused myself with ice-cold water from a tiny hose at regular intervals. Beads of sweat appeared across my body all at once. Twice I stuck my face out one of the porthole windows to breathe in the cold crisp air and survey the grounds. It was still quiet. Finally, I thought I may have had enough. I decided to rally to the cold dip.

Outside the sauna is a water hose that carries mountain stream water to a large antique claw-foot bathtub that was already partially filled: the cold dip. On past visits I had watched cheerful saunaers howl and quake as their counterparts threw heavy wooden buckets of water over their heads in the dip before scuttling back into the comparative comfort of the sauna. I had always resisted the cold dip, my desire to avoid doing anything unpleasant, even if only for an instant, outweighing any other benefits.

I stepped gingerly into the tub and began by splashing some water on my legs and arms before finally giving in and thoroughly dousing myself with the hose. I gasped in spite of myself, startled at the very coldness of the situation. When I looked up, I was standing face to face with a mature deer. She was less than ten feet away, staring intently at me, naked and red-bellied. We stood in stillness together for minutes, until the steam failed to rise from my skin. I turned to go as she also slowly moved on, her head still turned towards me, even as she made her way down the trail and out into the woods.

Strange coincidences are common at Breitenbush. It is just as likely to meet one's future acupuncturist in a pool as it is to run into one's accountant or grade school teacher. Ten years after my morning with the deer, a woman who it turned out had worked for my father two decades earlier turned to me and asked, "Have you heard about the woman who found the child in the woods?"

I had.

Two days earlier, before checking in, I parked my car along a forest service road and headed out north on the Spotted Owl Trail, hoping to find the connection to the Pacific Crest Trail. An hour or so later, I was stopped at a poorly labeled junction, unsure of which way to go when he appeared. A boy, moving fast around the bend, then pulling up suddenly when he saw me. I smiled at him and turned away, assuming his parents would appear shortly.

There was a loaded silence at my back. I looked back around, and he was there, now asking if I had seen a dog. I told him that I hadn't and turned back to the sign. But alarm bells were going off in my head. He looked a little wild-eyed and scared; he had lost his dog, and he was dirty. The kind of dirty that doesn't come from being outside, but from already having been dirty. My gut told me his parents weren't coming around the corner.

For one moment I became a terrible human being. In that moment I thought only of myself. I lamented my vacation, my day, my hike, my solitude. For just that one moment I thought to myself, *I don't want to do this. I just want to relax.* Internally, I railed at the universe for dumping this problem on me. I turned to him and asked, "Are you lost?"

"Yeah."

"Ok." I took a breath, and made a choice. "Listen kid, you've pretty much hit a gold mine with who you managed to run into out here. I'm a teacher and a camp counselor and I work with kids all the time outside. And I'm pretty much the safest person you could have found. But, since you're a kid, and I'm an adult, and I've found you, you kind of have to stick with me until we get you back to your parents, okay?"

He looked at me with his head tipped to one side for a very long time, which I figured was fair enough.

"Okay," he said.

Growing up in Oregon in the 1980s there were really only three big news stories: the Cold War, the Challenger disaster, and the Spotted Owl. None of which I was able to properly wrap my head around. To my grade-school self, the least scary of these were the owls, an issue which seemed pretty cut and dry to me. They were owls. They were wise. Also they were interesting looking. They deserved a tree to live in. Trees were nice too, and seemed important if you liked forests. Which I did. At school, amongst my peers, this was a surprisingly unpopular opinion. It seemed many, many families made their money from logging the owl's trees.

Industrial logging in Oregon began nearly two hundred years ago. In 1900 it constituted a major portion of the state's economy. By the housing booms of the 1950s and 1960s, more than twenty percent of all the timber, including nearly half the plywood, in the United States was harvested from Oregon's forests. Oregon was then, as it is now, the largest source of timber in the country. And at more than two hundred dollars per thousand board feet, and close to two billion board feet being harvested annually, Oregon's economy soared. In the 1980s, though, inflation and high interest rates squeezed the market, making the already difficult and costly process of downing and transporting trees that could exceed three hundred feet in height and ten feet in diameter nearly cost-prohibitive. Add to that delays and expenditures due to court battles over the spotted owl, protests against logging, and dwindling access to public lands…Oregon's economy ground to a halt.

So if you were for the owls, you were pretty much taking food out of your classmates' mouths. I stood my ground with a child's logic: "We're going to have to stop cutting down trees sometime soon anyway, if we want there to be any left."

Soon, I began to have a consciousness of the changing nature of Oregon's forests. The long drive to the beach from Portland, once shrouded in darkness by dense stands of trees, shone through with light, though the view of the clear-cuts was still well hidden by a thick screen of trees.

Trying to describe what it looks like when a five-hundred-year-old community of living things is decimated by clear-cutting is like trying to describe the vastness of the ocean or, rather, the absence of the ocean. Oregon's temperate forests are home to a startling number of living things, among them some of the tallest and oldest trees in the world. They are dominated by Western Hemlock, Douglas fir, Western Red Cedar, White Pine, along with more than a dozen other evergreens, species that arrived in the region as early as 50 million years ago but whose ancestors are preserved in fossils that date back to at least 350 million years ago. Over time they have adapted to become uniquely suited to western Oregon's cool, damp climate. The needles, which they retain all year, allow them to extend their growing season and conserve their annual energy expenditure. From a distance these trees obscure all other forms of life.

But portions of Oregon's forests contain as much or more biota per square foot than tropical rain forests, some of which is almost invisible. At the ground level, Oregon forests host myriad mushrooms, the aboveground fruits of a vast network of fungi, many of which live symbiotically with the trees. Forests grow like this, in mutually beneficial layers called sere with groundcover like mosses and lichen resting at the feet of an herbaceous layer of flowers and ferns, which in turn settle under shrubs of vine maple, salmonberry, and rhododendron mixed among larger trees of the understory, willows, big-leaf maples, all housed under a coniferous canopy that nearly blocks out the light.

In a clear-cut, all of this is dispatched with efficiency by a few dozen humans helming an array of several-ton heavy machines that do exactly as their names imply: fellers, harvesters, heavy-utility helicopters, delimbers and debarkers, grinders, and brush-piling shovers reduce the forest to bare, sun-cooked mounds of exposed soil and woody slash piles. Even the shrubs are sent through the chipper.

By the 1990s, all of the trees, even those comprising the roadside screen, would be harvested. Great scars of clear-cuts became common, and the fight for the trees was no longer about owls. Instead, there were lockdowns, tree sits, and demonstrations on the streets as activists fought to preserve the forests. The news showed nightly images of loggers dragging protesters by their throats to open road gates and nearly running them down with tractors. There was also a great new wave of red tape and regulatory activity: riparian zones, sediment counts, and landslide analyses. Proof, it was said, of the government's commitment to the preservation of our resources. The forest fell. Breitenbush endured.

The kid was still looking at me. Try as I might though, I couldn't get my mind to formulate a course of action. My feet rooted to the trail junction. My thoughts drifted.

Around the time I first remember hearing about the spotted owl, my family got lost while hiking in the Oregon forests. Long past our anticipated return time, my parents' demeanor changed. They slowed. The tone of their voices became more serious. My sister and I were hot and tired and getting crabby in a sustained way. My parents probably were too. Finally, we came to a stop. And there they were, lost, with darkness falling and two very tired girls. They discussed. They couldn't be that far off track; we hadn't walked very far. They could still see the ridge we had descended, and the sun was setting in the west; they had a sense of where we were. They were young and resourceful and figured worst-case scenario they would be fine overnight with us if we got caught in the dark; we all had a warm layer tied around our waists. It would be an adventure, a story to tell. And also quite likely entirely unpleasant. My dad thought we could backtrack and either discover where we had gone wrong or return the full length of the way we came before it got too dark. There was also the possibility that we were only a short distance from a connecting trail, or even the trailhead itself, it simply being longer than we had anticipated. But we knew how to get back the way we came. They turned around.

This terrified me. I was little, with no sense of time or distance with which to put the situation into perspective. I was still afraid of the dark, and I remembered the view from the top of the trail; it seemed like a really big place to be lost in. I was imagining lions and tigers and bears and Yeti and rodents of unusual size. I was thinking that this very kind of thing was what had happened to those Hobbits. My sister was fixated on spending the night out there; she may have said something about having to eat me first. In all, this probably happened over the course of just a couple of hours, but in child's time we wandered forever, like Lawrence of Arabia, parched for water and slowly weakening with hunger. Walking with my dad holding my hand, I started to cry.

It was at that point that I remembered, looking at the boy, that I also was not so far from lost. But I was sure I knew how to get back to Breitenbush. It had taken me over an hour of fast hiking to get where I was, so I knew it would be a while before we made it back. I offered him a granola bar and some water, neither of which he wanted, and we set off together.

He was nine years old, spunky, polite, and, like me that day in the woods with my parents, really scared. He was also dehydrated, disoriented, and entirely unprepared to be outside by himself. Or anywhere else, it seemed. His tennis shoes were worn through the bottom and, he told me, he had never been to school.

He was worried his father would be angry with him. He was supposed to be taking the dog on a short walk down a nearby trail, but the dog bolted and he ran after it farther and farther into the forest until he was lost. That he was smart enough to call out to me and ask for help was to his credit, but that was where his ability to help me, or himself, get him back to his family ended. He couldn't tell me where he was staying, what kind of vehicle his family drove, or even a cell phone number for his dad.

I asked about his mom, and he told me that he was angry at her.

"Oh," I said. "Why is that?"

"Because," he replied, "she had a choice between going to rehab and staying here with me or moving to Florida with her boyfriend, and she chose him." This last bit was emphasized by the sound of a rock narrowly missing a chipmunk and bouncing off a nearby log. He had been trying to hit one for days, he explained. I decided to let both his mother and the chipmunk fend for themselves, imagining the clear-cut of this kid's life: food deserts, educational wastelands, and the concrete of institutional walls.

Eventually, we made it back to the Breitenbush office. I stood in line, shifting nervously from foot to foot, not sure after such a long time if this was still considered an emergency. The hippie behind the counter looked at me, and then turned his head, to pointedly ignore me, clearly indicating that my sense of urgency was totally ruining the vibe. It took me three tries at telling him my story before he wrapped his head around it.

"You mean that this kid isn't with you?"

"No. He's lost. I found him, and he's been gone for at least a couple of hours, so if you could start making calls that would be great." I figured they had a system for this, that once he picked up the phone people would descend, taking the child from me, perhaps to the lodge, giving him a snack, and entertaining him until the proper authorities or his father came to claim him. But no.

"Call who?" The kid behind the desk was still sitting motionless, his hand on the phone. I looked around at the other adults, all of whom had backed away from me so as to not get any of the problem on them. They too, looked like they didn't want this kid to ruin their vacation.

"The Sheriff."

"What's the number?" And here, I almost lost it. That the front desk of what had immediately been reframed in my mind from a rustic retreat to essentially a white-privilege new-age resort in the middle of the woods didn't keep a list of emergency numbers at the front desk, or have any kind of emergency contingency planning stunned me into temporary silence. A rant started to form in my head.

I didn't know at the time that it wasn't their fault that there wasn't an emergency plan in place. Decades of protests and lawsuits had alienated the community from the local government in an area that is still largely lawless. It makes for a tenuous balance. Here, rednecks that lose their children don't ask questions about the naked hippies up the road. And the hippies, they're supposed to keep to themselves. They don't have the Sheriff's number because it doesn't occur to them to turn to them for help. Are they inaccessible or do they just not care? All I know is that even after we called, no one came that day, and no one followed up.

I was held in check by the kid totally freaking out at the mention of the Sheriff. He grabbed my arm and looked up at me pleadingly, saying, "Please don't call the Sheriff. That's gonna make my dad so mad."

"Don't worry," I told him, even though I now was. "It's not that kind of call to the Sheriff, it's just that there's no one else to call out here, and he's probably already looking for you." But after fifteen more minutes of cajoling and prodding the front desk hippie, it turned out, he wasn't. This was the first the Sheriff had heard of a missing child. Did he know his Dad's phone number? No. Address? Campsite? How to spell his last name? No. No. No. He had never been to school. They would put out a bulletin.

There was a collective sigh of relief. And then everyone returned to what they were doing. The kid and I looked at each other. With no further offers of assistance and still unable to check in to Breitenbush, we went back outside the hut and sat on the only available bench, the backs of our legs cooking to it in the late afternoon sun.

After about ten minutes, we got bored. So I started over from the beginning. How long have you been out here with your family? Were you staying in tents or a cabin? Did your site have a name? A number? Was it near a paved road? The river?

I had asked all of these questions before, but time, blood sugar, and the growing fear that I might be with this kid in a semipermanent way, finally started putting the pieces together. I went back in to question the hippie and came out armed with a plan. There was a series of cabins and campgrounds not far down one of the forest service roads. I'd start walking with the kid and see if anything rang a bell.

Sure enough, thirty minutes down the road, a circa-1984 station wagon that showed its miles through the dust slowed in front us. Inside was a massive, gray-haired woman with an open can of beer in one hand, a cigarette hanging from her mouth, and plastic-rimmed glasses so covered by dirt you could barely make out her eyes. I asked the kid if he knew her.

"Yeah. That's my grandma."

The car rolled to a stop.

The dust-slathered window slowly rolled halfway down and her glasses peered out at me like giant raccoon eyes in the sun. It was clear she wasn't getting out. It was also clear that the kid had been right, she was pretty pissed off. And suddenly, I didn't want to let him go.

He's in and buckling his seatbelt, for what good that might do him in the face of what life has already handed him. The car immediately begins to roll away and I lose the chance to offer him any words of advice, or even say goodbye. So I stand and watch him go, his head turning to look at me as they head away from Breitenbush and into the unprotected woods.

FIVE CENTS A BUNNY

ET ME TELL YOU A STORY ABOUT RABBITS. THOUGH IN truth, it isn't really my story to tell. The owner of this particular story, an artist, took his own life at the height of the Great Recession, a casualty of those strange and trying times. It was an incident that haunted him, filling his art in the following years with images of rabbits, some magical, some terrifying, often large, mounted across the walls of his living room. His name was Hans, and I don't think he would mind.

It happened late at night on a dark Oregon road in the middle of winter, most likely someplace in the coast range, which was a common place for him to drive, aimlessly, until he reached some overlook or other along the coast. On this night, Hans had the narrow, winding road to himself and had allowed his mind to wander when he saw something flash across the headlights of the car. Even though he hadn't felt a bump, something told him to pull over. When he stopped, he could see an object in the road in the review mirror. He got out, walking several hundred yards in biting cold without a coat, until the shape took form. It was, as he described it, a massive white rabbit.

As he moved towards it, a second rabbit appeared out of a tangled wall of Himalayan blackberries, snowberries, and wild roses along the road to stand and sniff at his companion's unmoving body, enveloping it in its arms. It stayed there, Hans recounted, for minutes, cradling its wounded friend. Eventually, it broke the embrace and began to pull the lifeless body from the road. The thing of it was, he told me, that rabbit, it was screaming—a loud, high pitched wail that he could hear even after the two had moved off the road and out of sight.

If ever there were an argument for the rabbit's possession of a soul, Hans's story is it.

Rabbits and their cohorts, hares and pica, evolved over 40 million years ago, long before the first true humans walked the Earth. They are found in every part of the world except Antarctica, yet in spite of their geographic vigor, rabbits are short-lived. Nearly eighty-five percent of wild rabbits live less than a year. Rabbits and hares, while used somewhat interchangeably in the common vernacular, differ in several ways. The most distinct difference is that only rabbits, not hares, burrow, forming vast underground networks called warrens. In contrast, hares build aboveground nests called forms, usually under dense brush. The National Audubon Society's Guide to the Pacific Northwest lists six species of rabbits and hares, four of which are found only in the deserts of the eastern portion of the region. Of the remaining species, only one could be described as "a massive white rabbit."

The snowshoe hare is described as nineteen inches long, over a foot and a half tall, and topping four pounds on average, but has been observed to grow to over eight pounds. They are nocturnal, inhabiting the dense underbrush of coniferous and mixed forests and commonly feed along human pathways such as trails and roads. They are popular prey for a wide variety of predators, so much so that lynx populations in Oregon match their booms and busts. As a result of this, they have developed an advanced system of seasonal camouflage, turning from dusty brown to white in the winter. It is a strategy that works: the average jackrabbit, which utilizes a similar camouflage technique, lives nearly two years longer than smaller breeds of wild rabbits. Even with such protections, however, they are a nervous bunch, prone to running out of fear at speeds of up to thirty-five miles per hour along bolt paths, previously established escape routes to their safe rooms in the brush. And there's this: While most species of rabbit and hare are silent, communicating by thumping their hind feet on the ground, the snowshoe hare, when distressed, screams.

Their fears are not unwarranted. As prey goes, rabbits are pretty low on the food chain. Their predators include nearly every kind of meat eater from bobcats, cougar, and coyote to birds of prey, snakes, and even weasels. Nothing though, compares to the human penchant for the rabbit. Humans have hunted rabbits throughout our history and have raised rabbits for meat in a practice called *cuniculture* since the Middle Ages. In modern times, rabbits are domesticated as pets and bred for laboratory experiments and product testing in addition to being raised for their fur, which is used in a variety of textiles. In culinary traditions, they were bred for cuisine as early as 1,400 years ago by French monks because the Catholic Church allowed them to be eaten during lent.

In the United States, wild rabbits used to provide more meat for American tables than any other game. The abundance of rabbit preparations and treatments in mid-twentieth-century recipe books illustrate this: including rabbit casseroles, poached rabbit, baked and stuffed rabbit, buttermilk fried rabbit, and the traditional *hasenpfeffer* (rabbit-pepper) stew. The 1953 edition of the *Joy of Cooking* offers a basic recipe for this mild-flavored stew. They recommend searing and seasoning your rabbit before dousing it with a jigger of brandy, lighting it, and allowing it to burn. Then simply place your rabbit in a pot with onions or scallions, pepper, and other savory seasonings, cover it with a one-to-one ration of stock to wine, and cook, until tender.

Historically, and to this day, rabbit hunting in Oregon has required no tag and has had no season or bagging limit. As such, it remains a popular off-season hunt, especially in more rural areas to the east, such as Harney County.

Harney is one of the largest counties in the country, nearly the same area as the entire state of Maryland and, with a population of only 7,700, cows outnumber humans fourteen to one. Nevertheless, it became the focus of national attention in 2016 when a group of armed militants seized the Malheur National Wildlife Refuge offices in a bid to force the Federal government to hand over the refuge and other federally managed lands to the states. Dozens of militants held the offices for five weeks, the standoff eventually resulting in surrender and one death. During the occupation, though, they dug new roads, spent weeks using sensitive Burns Paiute Tribe burial sites as latrines, and did so much damage to the facilities, files, programs, and infrastructure that efforts to curb the population boom of the invasive carp in Malheur Lake were set back an estimated three years.

In 1987 Catlow Valley homesteader Raymond Olsen was interviewed about life in Harney County, Oregon. Casting back to the early twentieth century, Olsen described the land at that time as being riddled with rabbits—jackrabbits, cottontails, pygmy rabbits—all coming out of the brush at night to eat the crops "to nubs," the newly planted alfalfa being a favorite. The problem worsened as increasing portions of the desert lands of Oregon were irrigated and planted with leafy crops, and coyote populations decreased due to hunting and land privation. By 1912 Oregon was overrun with rabbits. After being petitioned by the residents, Harney County finally took action, levying a tax to fund the institution of a bunny bounty.

At just eight years old, Olsen set snare lines along his route to school and checked them twice a day, killing and scalping any rabbits that he caught and stringing the ears on a wire loop that he would turn in at the nearest store for the price of five cents per pair, which he would use to purchase bacon, prunes, salt, flour, and ammunition. The county paid a bounty for predators like coyote and cougar as well, but rabbits were harmless and in abundance. And at five cents a bunny, industrious men and boys could earn a dollar a day to supplement their income— good money for folk still living in walled tents in barren desert country. More than 156,000 rabbits were killed in the first six weeks of the program.

This was not the first approach to controlling the rabbit population in Oregon.

The first approach was the rabbit roundup.

Rabbit roundups, or drives, as they were often called, were prevalent throughout the American West well into the twentieth century. They consisted of "rousing rabbits" with hounds or by beating the brush with sticks and then driving the rabbits into low enclosures or nets before allowing paying customers to club them over the head. At times more than 12,000 rabbits would be rousted, chased, corralled, and beaten to death at days-long events that seemed more like socials than work parties. During the day women served coffee, tea, and sandwiches, and they were typically followed in the evening by a *hassenpfeffer* feed. Pictures from these events show cheerful men and women in suits and lace-edged, high-collared dresses, posing proudly next to piles of rabbits that reached over their heads.

The organizers from one such event in Southern Oregon acknowledged the possibility that 'killing harmless bunnies might seem inhumane," but asserted that it "is astonishing how quick pity subsides" when one witnesses the destruction the "bunny" hands down to a young orchard or vineyard. That particular event was viewed as a success as the death count of jackrabbits was 1,200. From 1875 to 1896 there were 305 official rabbit drives throughout the west which together killed an estimated 635,546 rabbits.

In spite of these astonishing numbers, roundups had little to no impact on the overall rabbit population. As early as 1915 the *Oregon Sportsman* called them "a good example of over-throwing the balance of nature." But the roundups continued because they seemed to serve a more important purpose—allowing urban Oregonians to blow off steam. One enthusiastic participant underscored the point, noting, "Everyone felt good about it except the jack."

Ultimately, the Oregon State Extension Service took pity on the rabbits and convinced farmers to switch to a pest-control solution: poisoned alfalfa.

In the twenty-first century, rabbit roundups and cuniculture have fallen out of favor as the agri-industrial complex has removed the population from its food sources and narrowed the choices and palettes of the American public. In 2003 the Oregon Department of Fish and Wildlife Services canceled its own event, a family rabbit-hunt day that was part of their ongoing Becoming an Outdoor Woman program, which teaches outdoor skills and safety. Animal rights advocates had objected, saying that it encouraged cruelty to animals and young people to play with guns. Twenty-six people had registered to attend.

Rabbit populations, however, are not booming as a result of this shift in compassion and policy. Habitat destruction, wildfires, over-hunting, and decades of poisoned alfalfa have severely constricted population growth. As a result, some species, most notably the tiny pygmy rabbit, are now listed as endangered.

All this is happening just as economists, climate change experts, and food scientists are beginning to argue that rabbits, of all creatures, could save the world.

Rabbit meat is leaner than most other meats, making it a healthier choice in this era of first-world obesity and declines in life expectancy. Furthermore, rabbit meat is allowed by Muslim law, kosher, and considered acceptable for consumption by Hindus, making it one of the most widely accepted meats in the world. Moreover, rabbits do not feed on grain but on their associated leafy greens and surrounding grasses. Replacing beef and other forms of meat with rabbit, which reproduce at significantly higher rates than other animals bred for meat, would greatly decrease the impact of producing meat proteins on the world food supply, enabling significant amounts of grain to be transported directly to the mouths of people. Since rabbits are found throughout the world, switching to a rabbit-based diet would encourage localized production (in turn increasing the stability of the global food chain), encourage local economies and related industries, and greatly decrease the carbon footprint of agriculture at large.

Why then, do we not embrace a contemporary, rabbit-fueled, diet? Surely, in a modern world we must have the ability to see beyond cuteness and sentiment to issues of sustainability and nutritional justice. And yet, we hesitate, stymied by a reticence surely related to our perception of some form of humanity in these hunted and frightened animals.

Part of this is undoubtedly the oversize role the rabbit plays in our imaginations. Rabbits and their peculiar behavior have permeated secular culture and language for centuries. Take, for example, "Mad as a March hare," a popular British saying referring to the boxing behavior observed in hares in the early weeks of the mating season as over-aggressive males are beaten off by picky females. That analogy served as the basis for the March Hare character in Lewis Carroll's *Alice in Wonderland*, perhaps the most pervasive and enduring example of the rabbit as personified character. Fictional hares Bugs, Peter, and even Pooh Bear's Rabbit, are often portrayed as comical victims, set upon by villains, predators, and troublesome neighbors. Closer analysis, though, reveals a more complex pattern of portrayal in which the rabbit serves as a proxy for our own existential crises, the best example of which may be the Velveteen Rabbit's quest to become real.

It is a practice that extends far beyond children's literature. In folklore traditions worldwide, rabbits are symbols of springtime, good luck, and fertility. They are renowned and celebrated for their skittishness and fear and envied for their ability to disappear into the ground—the metaphorical netherworld. In stories from the early peoples of the central United States and Canada rabbits are "fear callers," entities that project their own fear, attracting the very predators they hope to avoid. Mythology often casts rabbits as symbols of sacrifice, and some, Catholicism and Buddhism most notably, have adopted the rabbit into their stories of new life and redemption.

There are others. Carroll's White Rabbit and its descent into the rabbit hole is a reflection of ancient Celtic beliefs. Before Christianization, the Irish practiced a form of earth-based polytheism that emphasized animism—the attribution of a soul to plants, animals, and natural phenomena. The Celtic hare played a central role in their belief of the presence of the divine in the material world. They believed that rabbits burrowed underground in order to better commune with the spirit world, that they could carry messages from the living to the dead and from humankind to the faeries.

Which brings me back to Hans. He wept the night he told me the story about the rabbits in the road and he wept on the several occasions he returned to the story. The rabbits seemed to call him to the underworld. For those of us around him the rabbits he depicted in his art came to symbolize the most damaged and calloused parts of humanity. Like the Portlanders with their clubs or a young boy with a flour sacks' worth of rabbit ears on a wire, the rabbits are a sign of our collective descent. And we are in descent. We face an urgent dilemma, a tangible existential crisis to which rabbits may be the answer if only we would act. Our continued refusal to do so can only prove one thing—we are all mad here.

WHAT THE
OCEAN TAKES

THE OCEAN IS A TREACHEROUS WONDER
On February 5, 2011, Jack Harnsongkram and Connor Ausland, two strong young men from Eugene, Oregon, drowned when a nearly waist-high sneaker wave swept them off a rock bridge into the narrow chasm 100 M north of here. The force of the freezing, churning water and the shape of the rocks made it impossible for Jack and Connor to climb to safety and equally impossible for their four friends to save them without losing their own lives.

Jack and Connor died within three minutes.

—COASTAL SAFETY MARKER, YACHATS, OREGON

YACHATS HAS BEEN OCCUPIED BY PEOPLE FOR AT LEAST a thousand years, though little of its history remains. For centuries before the arrival of Lewis and Clark, Native American tribes lived in the coastal forest as hunter-gatherers, migrating to the ocean from the inland valleys seasonally, marrying their lives to the rhythms of the natural world. They called the area Yahatc—dark water at the foot of the mountain—and lived there in concert with the sea, in low pit houses built directly into the rocks at the water's edge. When the first white settlers arrived to the Willamette Valley, they took the land and its resources by force, marching the peaceful tribes north and west to the ocean near Yachats. Those that didn't die of starvation, exposure, or exhaustion along the way were lost to disease or forced into internment camps. Many tried to escape and were plunged to their deaths in the icy waters or were beaten and buried alongside hundreds of others in mass graves along what is now Highway 101. Entire tribes were lost this way, along with any proof of their existence. Most prehistoric sites in the area have been destroyed by construction, vandals, and the persistent work of the water on the land. What the ocean hasn't taken, it has buried. The remaining sites have been slowly swallowed by the water's interminable rise.

In recent years, Yachats has cast itself as a coastal destination. There are bed and breakfasts and oceanside hotels and restaurants from which you can watch families make their way along the network of public hiking trails that hug the shoreline, beckoning visitors to the water. It is a place of easy leisure as the rocks call to beachgoers with tide pools filled with green-brown anemones, cone-capped limpets, and deep-purple starfish, and the ocean offers up spectacular displays of surf and spray. The beauty of the landscape and the constant change in light through the clouds and against the water encourage a kind of mindful thoughtlessness, an easy disregard of our immediate surroundings that puts us into needless danger. This ocean, the locals will tell you, takes children like stones.

Unlike the disappearance of the Native Americans, which was barely acknowledged by white people for over a hundred years, until the 2009 dedication of Amanda's Trail (named for a blind Native American woman who suffered greatly in the marches). The news of Jack and Connor traveled inland fast. I hadn't known the boys, but I did know Jack's mother, Sarah. We took dance together, and for a time, she was a predictable fixture in my life. She was an elegant and joyful dancer and though I never saw her outside the studio, she was a kind of friend. The Monday after the ocean took Jack and Conner, Sarah came back to class to dance with us for a tenuous two hours. She seemed glad to be there, following the rise and fall of the drummers across the floor, but after that day, she never came to dance with us again, her absence leaving a great black hole in the space between us.

Waves are the ocean's foot soldiers. They form as nearly constant winds sweep across the water, rolling it in great circles. As the sea shallows towards land, the base of each wave finds purchase on the ocean floor and is slowed by friction. The top of the roll continues unencumbered, outpacing its base and rising as a cresting wave in a process called "standing up." In Oregon the water is deep and there are few obstructions to slow incoming water as it approaches the particularly steep continental shelf. Waves stand up quickly in these waters.

After two years of dancing without Sarah, I stumble upon a video about the installation of a coastal safety marker where Jack and Connor drowned. The video talks about beach safety, telling their story and showing a scene from the dedication of the marker in which Sarah reads while a Native American woman plays the flute. Later, they film her as she stands facing the ocean, her grief like the surf, standing up quickly, as she watches the water pass under and over the narrow stone bridge from which the boys were swept. She looks into the ocean, her small, pale hands framing her face as she works to tuck lost strands behind her ears. I grieve too, to look at her. It's painful to see how deeply she has contemplated the scene. She begins to talk about the ocean, the rhythm of the waves, their irregular and syncopated beat, and about the missed warning signs of the water's danger. She points out how the rocks are covered in barnacles indicating they spend time submerged beneath the water, how frequently a larger wave will wash across the bridge, disappearing it entirely. How they might have known. The video shows a class picture taken an hour before Jack and Connor were swept away. In it they stand with their classmates, shoulder to shoulder, wide grins across their faces. Of the group, only Jack and Connor have their arms raised, uplifted to the sky, as if they might have known. The scene cuts back to Sarah, who is forgiving the ocean. "I still like it here," she says. "I come and listen."

It's not just the water that needs forgiveness. The rocks are also partly to blame. The very cliffs and crags that make the coastline beautiful have proven treacherous and for centuries have taken the lives of those who choose to test the waters. But it is the conspiracy of the water with the rocks that is the most dangerous, the more complicated the shape of the rock outcrops, the more unpredictable the surf becomes. As water bounces against the rocks the energy of the incoming waves is reflected and refracted, the troughs of some waves canceling out the crests of others, while still others combine, creating waves of unusual size and force, *sneaker waves*. Sneaker waves are thieves and thugs. The force of even a small sneaker wave is enough to lift you from your feet. The backwash from sneaker waves is just as powerful as the swash; not only do they knock you down, they drag you back out with them. Oregon's Pacific waters are characterized by these waves. Between the years 2000 and 2013, twenty-five people lost their lives to sneaker waves along the Oregon coast and at least twenty-six people were swept into the ocean but managed to survive. It seems to be as easy to be taken as it is to be left behind.

I go to Yachats often to sit and watch people venture out onto the rocks. I listen to the uneven rhythm of the water, think about Jack and Connor, and read the prominent stain-less-steel-and-stone safety marker that tells their story and the warning that follows it. There are few such markers on the Oregon Coast, so it is unusual that just four years before it was installed another, quieter marker was dedicated, deep in the woods along Amanda's Trail. It was a simple carved-stone statue of a Native American woman grown over with moss, meant to remind us of the historical truth of the place and the lives that were lost to coastal forces entirely within our control.

The ocean though, is indifferent and beyond our control. The dark waters of Yachats are especially so; they take houses, lives, lovers, and old couples equally, regardless of heritage or skin color. They have swallowed history, erased cultures, and buried secrets, and they continue to do so. On December 18,

2015, storms washed the bridge to Amanda's Trail and the statue into the sea. Now it's up to us to preserve the buried secrets of our past.

> *Respect the immense power of the ocean.*
> *Know the tide level.*
> *Face the ocean at all times.*
> *Enjoy the beauty of the coast safely. Speak up to others who may be in danger and to those taking risks along the shore, listen.*
> —COASTAL SAFETY MARKER, YACHATS, OREGON

SCABLANDS—
LOVE AND THE
MISSOULA FLOODS

A T THE TURN OF THE TWENTIETH CENTURY SCIENTISTS from several disciplines began to warn of the conse-quences of a rapidly accelerating aspect of climate change: the melting of polar ice. Significant warming, they cautioned, could send huge, landmass-sized portions of ice sheets into the oceans releasing encased water and displacing massive amounts of seawater. The effect on humanity, they concluded, would be the rapid and catastrophic rise of global sea levels by seventeen or more feet, enough to swallow cities forever and change life as we know it.

In 2017 NASA's Operation IceBridge, its comprehensive aerial survey of the by-then rapidly diminishing ice caps, photo-graphed a heart-shape portion of a glacier in northwest Greenland as it cleaved in two, proof of the predicted large-scale melting.

Between those same years I met the man that would become my husband and knew him for sixteen years before either of us realized we were in love. Like the melting of polar ice, it was a process that once begun, happened fast, taking less than a year to cleave us from our prior lives, nearly losing everything in a flood of punishing consequence and angry meltwaters.

This is the story of how catastrophe unfolds.

Nearly seventy percent of Earth is covered with water. It is perhaps the most anomalously contrarian of all substances, the best example of which is its stalwart insistence on being less dense as a solid than as a liquid. From a geologic viewpoint, however, it is just another rock—a mineral in fact, albeit with a melting temperature far below that of normal rock—but a rock all the same, subject to the same rules of nature as any other substance. Still, its polarity, unusual structure, and thermodynamic properties give it peculiar power over life and land. This is especially true in glaciers, where it acts as a slippery between-phase substance that feigns solidity while secretly, slowly, flowing across the landscape.

The Earth has spent a lot of time encased in ice. The rock record tells us there have been at least five ice ages in the 4.6 billion years of Earth's history. The one in which we currently find ourselves began 2.6 million years ago and it has waxed and waned several times, advancing glacial blue ice like bulldozers across continents and towards the equator until variations in the Earth's atmosphere and axial orientation create enough warmth to force its retreat. The landscape of the Pacific Northwest is a record of one of the most rapid and large-scale of these warming periods and its resultant catastrophic events in recent geologic time—The Missoula Floods.

The human heart, like Earth, spends a lot of time encased in ice. Never mind that it, too, is composed of nearly seventy percent water, the ice that holds the heart is pure fiction. It is our armor, the emotional and psychological barrier we build between our inner selves and the outer world. That cold shell we construct protects not just our romantic loves but our joie de vivre, childish delights, sense of awe and wonder, and most intimate hopes and dreams. We lock these warm, tender things away behind a cover of ice in the name of conformity, for fear of being condemned, ostracized, or ridiculed. Often it is our most important passions, loves, and desires that are suppressed and denied. It's a dance between our true selves and society's expectations, a surrender to social constructs determined by religion and policed by politics.

The landscape of my heart at my mid-thirties was scabbed and scoured from a procession of unhappy relationships. They were unions forged out of youth, circumstance, and a vague desire to marry, nest, and have children. Like many relationships, they were contracts, designed to fit the participants into the model of who and what a marriage, or a woman, should be. But differences—of paths, ambitions, and value systems—had seemingly always paired with bad taste and worse luck. I kept moving on, afraid of being trapped in the kind of quiet desperation I saw in the lives of people around me who were drowning in compromise and obligation. But dashed hopes, and at least one hard miscarriage, had left me with scar tissue.

The summer Paul and I fell in love we were buried beneath the ice. Both of us were partnered, with jobs, commitments, and in his case, kids. From the outside we each approximated mainstream normalcy, but on the inside, neither of us were happy. We didn't know this about each other, or very much else for that matter. Because up until then, our entire friendship had existed entirely within the confines of a burger booth at an annual three-day summer festival, the Oregon Country Fair.

"Fair" is a close cousin of Burning Man and widely considered the granddaddy of all summer arts festivals. Forged out of the back to the land movement and propelled by early Grateful Dead shows and Ken Kesey and his Merry Pranksters, Fair began in 1969 in Eugene, Oregon, as a Renaissance Festival and quickly became a multigenerational cultural phenomenon filled with whimsy, absurdity, and old-time arts, dance, and music. It's a rabbit hole of belly dancers, acrobats, light displays, giant bubbles, discos on wheels, fire dancers, and marching bands with an abundance of wildly costumed, smiling people, many nude, often traveling by stilts and unicycles or floating past as illuminated jellyfish, life-sized puppet trees, or red-rumped pigs. It's a place to escape the confines of daily life, indulge in playful revelry and discovery, and expose yourself to a cultural kaleidoscope. At Fair, it is just as common to spend an hour Hula-Hooping as it is to find yourself making a smoothie with a bicycle-powered blender or engrossed in conversation with a besparkled stranger body-painted to resemble a large, iridescent cat. It is a place where the unofficial motto is "Yes! Yes! Yes!" and no one who wanders is lost. It is life, vibrant and intensified.

More than that, Fair is a democratic response to the confines of mainstream existence. And it's been that way by design since the beginning. Cynthia Wooten, one of the founders, once described the initial vision to me as a response to the confining pressures of the time, a natural extension of women's rights, civil rights, and the "conservation awakening." It was, she said, a rejection of what they saw as a flawed American Dream. Fair was to be a demonstration of community and self-reliance, a means by which people could come together to sell things of their own making and talk about how to craft an alternative future.

Paul and I were worker bees, two of the twenty-some-odd-thousand volunteers and booth staff that kept the machine running in exchange for an overnight camping pass. It was an experience we both stumbled into as teenagers through the vast and complicated local network of underground businesses and social circles that drives the Fair and a large portion of the local economy. Neither of us is sure what year we first met, but we do know that because of the overstimulation of Fair and a constantly rotating roster of staff, it took us at least three years to learn each other's name.

When we finally did, it was because we were outliers, morning people among the partiers. Rare birds at such events. I worked the breakfast shift each morning cutting bucket after bucket of potatoes at a long wooden table at the back of the booth. He sat on a nearby cooler drinking his coffee and chatting. At eleven, when the public was finally let in, Paul took over the grill, frying burgers for masses of hungry hippies and the people who came to stare at them. It was hot and thankless work, but it paid our way in and the owners were generous about feeding the staff, an essential bonus in what could be an expensive sojourn. As the only vegetarian member of an oddly out-of-place burger joint that proudly and prominently displayed a "No Shirt, No Shoes, No Tofu" sign above the grill, Fair could make for hungry days. Paul took pity on me. After my shift I would take up his spot on the cooler and wait for the magical moment in the day when he had access to all the ingredients for a good grilled cheese sandwich: the bread from breakfast and tomatoes and cheese from lunch. Special order, without fail, for fifteen years.

The moment we finally fell in love it was over that potato table, in one look, held so long and so hard that it stopped a teenage boy passing through the kitchen in his tracks, wondering if something had gone wrong. To this day I can't remember what triggered it. But I do remember the way Paul staggered, nearly dropping his cup of coffee, and the way I held one hand to my throat, which had blushed so deeply it burned to the touch. From someplace inside my heart I felt a deep wrenching. And then the floodwaters rose in a dazzling and ferocious lifting of the roots and soil of my soul. A singular thought forced its way into my head. Maybe, after a life of trying to make it work with the wrong people for the wrong reasons, for the sake of making it work, the person I was supposed to be with had been there with me all along. *Maybe it was Paul.*

FIFTEEN THOUSAND YEARS AGO, NEAR WHAT IS NOW MISSOULA, Montana, a massive glacial ice dam, which had been helping to contain the meltwaters of a retreating continental ice sheet, failed. In the early stages of the flood the water ran as a dispersed sheet across Idaho and eastern Washington, advancing as a torrent of water so destructive and so powerful that it scoured the land for hundreds of miles to the west. With distance from the source, the water collected, following the existing channels of the landscape and converging, like any other water in the region, in the Columbia River basin, where its depth reached nearly one thousand feet. At Portland, where the Willamette

River meets the Columbia, the floodwaters split, some rushing out to the Pacific Ocean, the majority turning south and surging uphill, against the flow of the Willamette.

The Channeled Scablands of eastern Washington bore the brunt of the flood's erosional force and was scraped to bedrock so severely that today it is still wide-open land. It is largely undeveloped country, pocked with giant potholes so large they are only visible from the sky and dry falls over which no water has run since they were carved by the temporary rivers of icy meltwater. In other places, the land is striated with mega ripples up to fifty feet high—formed by water estimated to have been nearly two hundred feet deep and traveling at more than sixty miles per hour.

They are badlands, but at a massive scale.

In other places, like the Dakotas, badlands are a testament to what can be accomplished with the relentless application of interminable force. These badlands are carved out slowly, their stark buttes and basins and solemn hoodoos formed from the cumulative erosion of rock worked incrementally over eons by wind and water. Scablands are unique; they are formed from violent events that appear to be nearly instantaneous in geologic time.

AFTER THAT FIRST FIERY LOOK, WE TRIED TO GO ON ABOUT our day, potatoes, and burgers. One of us said something about betraying our cool exteriors. But everything was changed. Walking along the paths that last night of Fair, hands clasped, the world appeared like always and as it had never been. The sky bent down to caress the ground, the trees glimmered and juggled their leaves, the air wept gentle music and nothing breathed while we shared a kiss that had traveled so long to find its home.

Later that week I met him for lunch in a blueberry field near his house. It was the first time we ever saw each other outside the Fair. We sat there in silence, eventually falling asleep with our foreheads resting together only to awaken hours later to watch a blue moon cross the sky. Over the course of the next few days, we laid ourselves bare, relieved to finally testify to the sadness, frustrations, and loneliness of our lives. Each separation was a wrenching, a pulling apart of hearts and bones that seemed to have always been entwined. Unable to face lives of deception and believing that truth is always better than a fiction, we decided to break away.

NOTHING IS EVER REALLY CUT AND DRY. FOR INSTANCE, WHILE we conceptualize the Missoula Floods as a singular event, it was anything but. In truth, it was a slow-moving catastrophe comprised of many smaller floods called *jökulhlaups*. An Icelandic term, *jökulhlaups* are large, abrupt releases of water from sealed glacial lakes. These releases occur when interior water levels reach a critical point, exceeding the strength of the surrounding ice. Because these lakes often sit atop thick layers of ice, the force of the water upon release is particularly intense. However, irregularities in the internal structure and melt patterns of each glacier ensures that every *jökulhlaup* brings its own flavor of destruction. In all, it is estimated that the Missoula Floods consisted of more than twenty-five failure and flooding events over the course of approximately four thousand years. And even that is only part of the story. The Missoula Floods were just one of many flood sequences in a larger process of land formation and transformation due to the advance and retreat of ice that has occurred over and over again in the history of the Earth.

OUR PERSONAL *JÖKULHLAUP* OF THE HEART UNFOLDED IN much the same way.

Falling in love is not uncommon, or even very difficult. What is not easy is untangling oneself from your life. As we tried to figure what we were going to do, my social conditioning screamed at me. Tired, premillennial tropes about marriage, family, choice, and duty circled my mind causing me to question us. Sure, I thought, I could walk out of my relationship if I wanted, but there was no guarantee that Paul would be there. To do such a thing would take a herculean effort on both our parts and change the entire landscape of our lives. But once we awoke from the slumber, it was a matter of survival. And a few days later, walk out was exactly what I did, packing my belongings while my partner was out of town. We passed like ships in the night. Thirty minutes after pulling away, I got a true-to-character emotionless text that only served to reinforce my decision—"I guess that's it then." He never spoke to me directly again.

I holed up in a friend's extra bedroom, waited for Paul, and tried to be brave. And a few days later Paul did, in fact, come for me.

The short term was brutal. Judgment came in a chorus of righteous voices from friends, family, casual acquaintances, and even people we had never met, saying that while they understood that we may have been unhappy in our old relationships, our new relationship was doomed. They had seen our misery, our struggle to make it work with our previous partners, with their own eyes. But conformity, we found, is king. We were not, we were regularly informed, allowed to do this. And, in spite of any nonconforming decisions—or even transgressions—in their own lives, they meant it. As if, by that point, going back was even an option.

The suddenness of it all and the predictable and ongoing conflict with our exes meant we had to leave many of our possessions (and some of our friendships) behind. We had no pots and pans, no kitchen table, new commutes to work, new neighbors, new banks and grocery stores, and very little money. Life was so chaotic in the first few weeks we lived together that I got disoriented one day on the way home and had to circle the neighborhood, unsure of which street I now lived on.

We survived though, and married one year to the day after that long, fiery look, at that same burger booth we had circled around for so long at the Fair. We worked one more year of Fair after that before giving it up for good, content that we had both found what we had been looking for all those years.

THE END RESULT OF THE MISSOULA FLOODS WAS, ON THE ONE hand, loss. On the other hand, it was the laying down of a new foundational layer. On top of the scour features of the scablands and into the Willamette Valley were deposited great piles of glacial and slack water sediment, rich with material from the past, pieces from all the lands over which the glacier laid and the floodwaters traveled. Those thick layers of sediment developed into a rich topsoil that came to sustain grasslands, oak savanna, wetlands and riparian forests. Today it is some of the most diversely productive agricultural soil in the world. Willamette Valley farmland alone produces 170 crops ranging from grass seed to tree fruits and nuts to grapes, hops, and cannabis. It is on those fertile soils that Paul and I have made our home.

After a period that seemed like both a lifetime and no time at all, we had built a life. We began to grow enough in our garden each year to make it through the winter with homegrown fruits and veggies. We were friendly with the neighbors and knew the local grocers. I stopped getting lost on my way home. Neither of us is very social, but we kept a small group of friends that stuck with us and some that we found along the way and we do not miss the friends we lost, which were not many. My parents and extended family adopted Paul and the kids to our clan with open arms while I remain a bimbo and a harlot to his—we didn't win all our battles. Our marriage is a happy one.

In geology, there is a guiding principle for interpreting the events that have shaped a landscape. It is a kind of doctrine of uniformity developed during the Scottish enlightenment of the eighteenth century and simply articulated by Charles Lyell as, "The past is the key to the future." As the underlying assumption of all geology, it holds that the natural laws and processes that have always operated on the Earth are still at work today. That is to say, we assume that processes, and therefore landscapes, will conform to our notions of how they should be. To understand, then, how a place came to be in its current configuration, one would look for evidence of similar events in the past, a reference marker of sorts.

Sometimes, there isn't one. When geologists put Dave Johnston atop a nearby ridge to observe Mount St. Helens no one had ever seen or imagined a lateral blast, though deposits from such events are common around the world. But those deposits had yet to be explained and it took the eruption of May 18, 1980, and the loss of Johnston, to understand how a horseshoe-shaped volcanic cone could come to be. The exposed scablands of the Drumheller Channels in southeast Washington, where the Missoula floodwaters raged most fiercely, is another of those landscapes. The scablands record a fundamental experience of the landscape, their rugged contours elucidate new processes, new ways of being that have allowed the land to reinvent itself. These places serve as models for outlier events, those things that we have until now not been able to imagine how they came to be.

Happiness in modern life is also an outlier event. Although, Paul and I have learned, it changes everything, even if you have to lose everything first. It is not something that can be prescribed by a particular doctrine or system of belief. Happiness is there, amongst the layers of our experience, waiting to be exposed. Love is one path to happiness. Good love, love that chooses you, leaves room for adaptation, for the crafting of an intentional life, for friendship and equality in partnership. Things that don't often exist in bonds formed out of lust or a sense of how life should be. That kind of love reveals a happiness that isn't the hard work that people claim relationships should be and isn't made up of thrills or highs and lows, but is a quiet kind of contentment that arises out of a relationship whose participants are at peace with one another. And because of that, it's sustainable and expansive.

I'm still not sure exactly what happened that day at the potato table, but I do know that I wouldn't go back and do it differently. Years later, Paul and I hold steady to each other and our faith in knowing who we are as individuals and as a couple. And I know I can always be brave and he will always be there.

We cannot say that of the ice.

Since the Missoula floods the courtship between land and ice has remained relatively stable. At least, that is, until the last hundred years or so. Now humans have discovered that, to our own detriment, we play a role in that dance and that our choices are tipping the balance. Change is coming. The dawning notion of our ability to hasten these events, which otherwise might unfold interminably over eons, has been met with a feverish anger from those who deny that change has come or is even possible. It's a reaction that underscores our need for understanding one of our most basic human conditions, as subjects to forces and processes greater than ourselves. When the ice dam breaks, we will all be responsible for keeping ourselves above the water.

THE NECESSITY
OF TOTALITY

O REGONIANS LOVE ANYTHING THEY CAN USE AS PROOF
that they're the chosen people."
My sister texts this to me in early August 2017 in
response to my mention of the impending solar eclipse, during
which the path of totality, a sixty-mile-wide band in which the
total eclipse of the sun would be visible, was set to traverse
Oregon, our home state.

And she is right.

The hysteria had been building for several weeks. It began
as a low hum of questions everywhere you went—"What are
you doing for the eclipse?"—that had steadily ramped up into
a singular fixation until a driving attitude that the eclipse, and
totality in particular, was an event not to be missed. *Get thee to
the path of totality at all cost had become the mantra.* And with that
came the inevitable greed of too-hungry, Great Recession–rav-
aged Oregon small towns. Hotels began canceling reservations
made before they realized the significance of the date, jacking
up the price of the rooms by three, four, five hundred dollars.

When official campgrounds filled their reservation slots, desperate Umbrians began to buy campsites on private land, urban lawns, and municipal dog parks. In the far reaches of the state, places to which few Oregonians travel, miniscule airstrips prepared to land hundreds of small private planes. A Japanese businessman rented almost all of Sunriver resort, apparently flying in several hundred guests. By the time I was texting my sister two weeks before the event, every conceivable mode of commodification had been invented, from eclipse-themed mugs to doughnut displays. Some profiteers were even selling fake, sure to leave you blind, solar glasses.

A week before the eclipse the Governor declared a state of emergency citing the more than one million people estimated to arrive in Oregon, seemingly all at once. This stirred a new kind of panic, as the number was deemed far more than the state was able to handle. People began conducting mass balance calculations, detailed analyses of square footage of road space versus anticipated number of cars showing that Oregon's meager road system was unable to accommodate even a fraction of the anticipated guests.

None of this hoopla came as a surprise. Oregon has a long history of religious fervency and pilgrimage. The first pioneers to establish settlements in Oregon arrived in search of greater religious freedom. Drawn by the isolation and abundance of natural resources, the founders of the Willamette Mission and Aurora and Bethel colonies set the stage for more than a hundred years of spiritually based communal living. The term Holy Roller in Oregon references a small religious community largely regarded as a cult that formed in Corvallis in the late 1800s and was characterized by its participants rolling on the floor of their old, communal house in religious ecstasy. In the 1960s and '70s Jesus people arrived from all over the United Sates drawn by the promise of an alternative way of living fueled by cheap land and Ken Kesey's LSD. By the 1980s, the Rajneeshees were able to take over an entire town in eastern Oregon and poison several mainstream restaurants before anyone questioned their settlement.

The result of all this fervor over time has been the creation of a mainstream population prone to zealotry and a willingness to embrace nontraditional spirituality with passionate intensity. That is to say, this was not the first time Oregonians had turned a natural phenomenon into a quasi-religious circus.

In 1979 forty-one sperm whales beached themselves in the dunes south of Florence drawing hundreds of bystanders described by observers as drunks, screamers, prayers, and woo-woos. Instead of holding vigil for these slowly dying animals though, Oregonians climbed them to pose for pictures, cut out their flesh, and pulled their teeth in search of a memento. The next year, Oregonians took it on the road, traveling north when Mount St. Helens threatened to erupt, to stand as close to the evacuation zone as possible sporting volcano T-shirts printed in haste for the occasion, waiting for the moment the volcano would fail, for catastrophe, for excitement, for destruction. Or, as my sister put it, for proof of being the chosen people.

Eclipses are not rare. They occur whenever the Earth, sun, and moon lie in occultation, or direct alignment with one another. Occultations happen when these three celestial bodies all lay on the ecliptic, a great circle on the celestial sphere representing the sun's apparent path throughout the year and along which all the constellations of the zodiac lie. Still, they have always been deemed especially important astronomical events. Stonehenge, for example, constructed more than two millennia before the dawn of Christianity, is comprised of stone blocks and archways built using such complex geometrical relationships that it took years of investigation and ultimately the invention of modern computers for researchers to determine their purpose. The ring of stones is a series of sighting portals for tracking the eighteen-year cyclic nodal regression of the moon, the places in which the moon's path crosses the ecliptic. Stonehenge was designed for the prediction of eclipses.

Eclipse worship though, is not the sole domain of the religious. That these events hold deep significance for our understanding of the world is a widely accepted scientific fact. Einstein's Theory of Relativity was confirmed using measurements of stars made visible during an eclipse in 1919, proving that light from stars can be warped by gravitational forces. Even just the moon itself, whether it is in occultation or syzygy (opposition) to the sun looms large in the human psyche. Coastal communities and habitats live in close connection with the ebb and flow of moon-driven tidal waters, which, when paired with intense storms or rising sea level, become destructive or even deadly. Language though, might be the best indicator of the significance of Earth's largest satellite. Words like *lunate* (a crescent-shaped bone of the wrist), *empyrean* (considered the highest point of the heavens), and the more common *Monday* all draw on it for symbolism and names like Diana, Cynthia, and Phoebe, all related to goddesses of the moon, remain popular.

Total eclipses of the sun, while less common that other eclipses, still occur every two to four years but they are only visible from certain locations. In 2017 a path of totality had not crossed Oregon in one hundred years. The rarity of the event, dubbed the Great American Eclipse, seemed to generate excitement far greater than either the beached whales in 1979 or Mount St. Helens in 1980. Oregonians could barely be contained and were, it seemed, bent on illustrating the derivation of the word *lunacy*.

In the days before the eclipse the frenzy intensified as remote highways snarled with traffic when thousands of partiers flocked to a new-age pseudo-spiritual music festival called Symbiosis in a remote portion of the eastern desert that sat in the path of totality. Port-o-lets and temporary cell towers were trucked in from out of state to help handle the increased load on the rural infrastructure systems. The traffic jams stretched for hundreds of miles through a region where gas stations are scarce. Some people pushed their cars through the desert for fear of not having enough gas for the return trip. Bikes fared better on some highways than vehicles. In the more populated valley, hospitals set up triage tents on lawns. Fire drills were scheduled at manufacturing facilities to allow their employees to go outside and view the phenomenon. Classes for the schools that were in session were canceled for the day. Alcohol sales in the path of totality increased by forty-three percent.

This, twelve separate state agencies declared, was the canary. It could only get worse from here as more and more people pushed into the narrow band of totality. The state transportation department warned of days-long traffic jams after the event, fires started by viewers pulling their vehicles to the sides of the roads over dry summer grass, and accidents caused by people trying to drive while wearing their nearly blacked-out eclipse glasses. One official described the impending chaos this way, "When it happens, it will happen all at once."

And then, it didn't.

Streets throughout the state stayed empty. So too did stores, restaurants, hotels, and campgrounds. Staff at the Black Bear Diner in eastern Oregon occupied themselves charting maps of their few visitors' hometowns. News reports from coastal cities the Saturday before the event described normally bustling destination towns in the path of totality as spooky and deserted. By Sunday evening, the day before the eclipse, it became clear, the crowds were not arriving.

How could this be, when all evidence pointed to the contrary? Had we scared them off with our enthusiasm, our surety that people would choose us over any other swath of the country set to witness totality? Were our warnings about out-of-control forest fires, clogged roads, snakes, heat stroke, lack of gasoline, and coastal riptides too much? Had we overestimated people's desire to witness the event, blinded by our own fear of missing out? Was it that the frenzy in Oregon was not about the experience itself but some intangible element? Or could it be that purchasing the possibility of witnessing totality was more desirable than witnessing the event itself?

Perhaps so. Perhaps what we had witnessed over the preceding weeks was not so much religious fervency or a desire to connect with the divine but the physical manifestation of the modern desire for possession, commodification, completeness via consumption. It seemed that narrow stretch of completeness, of totality, had become a symbol of the American quest for fulfillment through acquisition and usance, of our desire to have it all.

Through all of this I was at home, looking up, from where I was, less than fifty miles from the path of totality, in a place where I would witness nearly 99.3 percent of the eclipse. A tragic near-miss, as Oregonians would have me believe. But how, I wondered, was that simply not enough? Even if it meant the ability to enjoy oneself and the event in tranquility or within one's economic means? To me, it meant that I would witness a singular version of the event, one that few others would get to see. And was there not something to be said for accepting exactly the portion of such a thing the universe saw fit to offer you?

So I held my ground. The hour before the eclipse was to start, I walked with my husband up to a small park on top of the hill by my house, joining perhaps two dozen neighbors armed with picnics and solar glasses.

At 9:04 a.m. on August 21, 2017, signs of the first total eclipse to cross the continental United States in seventy-nine years began to appear in the sky as the lower-left edge of the moon crossed the upper-right side of the sun. The warm summer morning began to cool. From my grassy perch I was able to see, for the first time, the three-dimensionality of the moon, and how very much closer it is to us than the sun. Shadows took on increasingly strange, cuspate designs. The world became still. Even the birds were silenced.

At the Symbiosis festival people prayed, cried, and hugged through the eclipse, unaware after days of drug-fueled party-ing and high-end glamping (an altogether twenty-first-century ecstasy) that they had not, in fact, become the chosen people.

Totality does not mark the end of an eclipse but the mid-point. Even so, back on our hill and in locations all along the path of totality, when that halfway point was reached and the moment of shadow culmination passed, all the people disappeared. They were in a hurry, presumably, to return to the dramas and pressing urgencies of daily life in a consumer culture, too busy after all to witness the event in its totality.

In defiance, we lingered, intent on seeing both halves of such a rare event, the whole of what was offered to us.

FIELD NOTES FROM THE DIGITAL FOREST

In the spring of 2019, I was awarded a writing residency at the H.J. Andrews Experimental Forest in the Central Cascades of Oregon, just fifty miles from my home. H.J. Andrews is a 16,000-acre ecological research forest that comprises one percent of the surrounding Willamette National Forest—one of the most heavily logged forests in the world. It is a place where clear-cut parcels sit next to stands of old growth. The residency, awarded by the Spring Creek Project, is part of the National Science Foundation's Long-Term Ecological Reflections Project. It is an effort to collect work from writers, artists, theologians, and philosophers of their impressions from five reflection plots throughout the forest.

The following are portions of my field notes from my stay at the Andrews, as recorded in real time.

APRIL 29, 2019, DAY ONE

THE DRIVE TO ANDREWS FROM MY HOME IS SHORT AND FAMIL-iar, on a winding highway along the McKenzie River. I drive slowly, passing farms and the diversion channel in Leaburg where the river is controlled; the tiny Heaven's Gate cabins that cling to the edge of the river's banks, asking to be swallowed by the next flood stage; and a clutch of popular riverside wedding lodges where I can count on both hands the friends I've known that married there, ticking off the unions that have since dissolved. Finally, I veer left just past the Christmas Treasures house, a local landmark where dozens of bearded Old Man Winters hold court.

It is a road I do not remember having taken before, though surely I must have. I have several memories of the Andrews, the first of which from some twenty years ago on a geomorphology class trip in the heat of summer to see the avalanche chute, a concrete plume that rises hundreds of feet up a steep slope and down which even today geologists chuck myriad types of debris to see what happens. In truth, the chute is a relic from a different time, rudimentary in conception and almost obscene in the way it cuts through the forest. It is also utterly unique in the world and unlikely to be duplicated—one would be hard-pressed to gain approval of such an invasive and industrial, albeit good fun, research project now. It was my memories of this strange, incongruous feature in the forest that drew me to apply to the residency.

On my way I stop at Blue River Reservoir, a popular summer recreation spot that is the gateway to the Andrews. I have it all to myself save a freshwater newt playing across the pebbles in the shallow shoreline water. The water is deep blue and glassy still, no bugs or fish disrupting its surface. The tree-covered mountains rise up sharply behind it and the world is still. It smells like home.

I arrive, check in, and go the "Green House," the only modern, non-forest-service brown or seventies-construction-style building at the headquarters. It has concrete floors, wood-slab furniture, and is clean and built to impress with large windows that look out to a graveled maintenance yard and repair shop and several other residences. Through them one might squint and think themselves in town.

In the afternoon I walk. The Andrews is a working forest and it shows. Besides the main building and residences there are the maintenance buildings, the monolithic avalanche chute, vehicle wash stations, picnic areas, and roads upon roads. In the forest there are stakes with thermometers, flagged trees, pipes and buckets, and everywhere freshly cut logs—a visible effort to clear roads and trails of the hundreds upon hundreds of trees that fell in a recent historic snow storm. They lay now in piles, already dry and seemingly waiting for a stray spark to begin their conflagration. I wonder if this is the last spring before it burns. Eventually, I pass the educational trail with its numbered markers. I was told that each child is given an iPad so they can watch an interpretive video at each location. I imagine rain-soaked groups of schoolchildren staring into screens at a virtual ranger. At each marker I stop and look, searching for the point of interest, but without an iPad I am at a loss. It could be anything, the fallen log, dog-tagged as part of some research project, the nurse log covered with a carpet of thick, fern-like moss, the ancient cedar tree towering above it all.

Back at the Green House, my every movement echoes against the concrete space. Outside military planes fly low overhead and there is a constant roar of chainsaws.

I RISE LONG BEFORE THE SUN, FILLING THE DARKEST MORNING hours with coffee and a book. A trio of green stones collected from the creek rest on the amber wood-slab table in front of me, begging observation. They are variegated, composed of angular fragments of parent rock broken and transported by landslides having sat through millennia of compaction, cementation, and erosion before being rolled, over and over, until they were smoothed and waiting for me at the edge of the water. I consider the value of practicing stillness like a stone.

Midmorning, Fred Swanson, a longtime research manager of the forest and eco-geologist arrives to orient me. I am pleased to meet this steward of the Andrews whose reputation as a researcher and protector of Northwest forests precedes him. He's a kind-eyed, lanky man who claims to be well into retirement but still sprints through the forest following what are only trails to his accustomed eyes. He is passionate about the Andrews and keen to talk about the research, which is abundant. With him as a guide, I notice for the first time the true extent of the research infrastructure in the forest. There are trees with wires and buckets running their length measuring temperature and humidity at ten-foot intervals into the canopy. Other trees wear belts, which measure their expansion due to heat, moisture, and growth. In some plots the infrastructure is composed primarily of white plastic (pipes, buckets, funnels) and when combined with lines for water, data, and power gives the scene a distinctly medical quality. I think this just as he points out a clunky thermometer mounting with a tiered apparatus attached to the front that says: "This is a respirator. It keeps the air around the thermometer moving so local variations in sunlight, for example, don't influence the data." Medical indeed.

Two hours and several miles in, Swanson is still talking about research. There are scientists of every ilk: the "owl people" who work at night, the "lichen people" who climb the trees to collect their samples. It goes on and on. For every kind of scientist, there is a research area. The Andrews is divided into three general types of areas: control plots, old-growth stands, and designated watersheds, all with studies being conducted in them. Geology dominates the research, which suits me fine. The substrate is volcanoclastic at lower elevations with dense lava flows and welded tuffs capping that at higher elevations. The pyroclastic deposits, with their loose ash and pumice are prone to failure, so there's a lot of geomorphology studies. The U.S. Geological Survey maintains the avalanche chute at headquarters and a series of stream gauges along Lookout Creek, a tributary of the Blue River, which itself flows into the McKenzie. In cooler eras, alpine glaciers helped form the crowded cluster of steep-walled valleys that make up the forest and their morainal debris had previously dammed the Blue River, which is now held in check by Corps of Engineers dams.

By the time we get to the scientists measuring soil creep across plots less than sixty feet square and the progression of a single tree being slowly split over years, one half on either side of a shifting slump block, the minutia of it all starts to overwhelm me despite my geologic background. I flash back to my own days as a researcher, entrenched in the singular-focus world of Newtonian investigation in which everything is parceled and compartmentalized. As I did then, I feel a frustration with the willful blindness of continuing these narrowly focused long-term projects, most of which were initiated when climate was considered reasonably constant. What is the point of all this long-term research in the age of rapid climate change? Steady-state, predictable systems no longer exist. That is, if the whole thing doesn't burn this summer. The question of fire sits unspoken on my lips all day.

At the old-growth reflection plot, I ask Swanson if outgoing research infrastructure is monitored as closely as incoming. Earlier in the day he had shown me a map of the research coded to indicate the purpose, location, and extent of each project to make sure that incoming investigators didn't disrupt existing plots. My question is about removal, especially after passing several sites of abandoned buckets and other remnants from clearly terminated projects. Abashedly, Swanson confirms that no, no one tracks the removal. Questions fill my mind at this. Is it like the Grand Canyon where debris left by researchers is "historical" by definition and must be left in place? Will future researchers arrive to study what these well-intentioned investigators leave behind? What could be learned from their buckets, wires, and netting decades from now? Will scientists use them to estimate rates of change, soil creep, or decomposition as they use trees now? Will they be buried under the weight of the forests' own debris or carried away by chipmunks or smashed by falling trees and deposited as fragments in the streams to be rounded like stones?

Then, in spite of the daylight, two owls hoot, emitting low, guttural barks between the trees. A mated pair calling out to each other, "Who cooks for you?" I never get a chance to ask. We return to the Green House in midafternoon. As I sit and write this, an oil tanker truck rolls past me for the second time today, refueling the forest. Earlier in the day I stood aside with Swanson as a UPS truck made its way past us down the road.

MAY 1, 2019, DAY THREE

I START THINKING ABOUT MONEY. IN THE HEADQUARTERS building, I ask the current lead of the site what he estimates the gross economy of the Andrews forest to be. Between research, managed logging, the Corps of Engineers, USGS, National Weather Service, Fish and Wildlife Services, and education and recreation, it's obvious that the Andrews is a forest with its own economy that surely is in the millions of dollars annually. The researchers clearly consider the Andrews a revenue source, the government, a sink. I am curious how this one percent of land compares to the larger Willamette National Forest in monetary terms and how that might play into future funding and protections. But no such accounting exists.

Then I ask if him he thinks this is a wild place. He struggles with the question, citing road density in particular as evidence of the touch of man on the land. But then, he says, "It's filled with wild things." The question hangs between us, unresolved.

MAY 2, 2019, DAY FOUR

IN THE MORNING, I VENTURE OUT DEEP INTO THE FOREST ON my own, eschewing the radio and check-in board, thinking I am heading to an easy hike, since from the map it appears to run level along the side of a spur. But elevation is deceptive in the Andrews. It takes me a full hour to cover slightly less than two miles on a narrow and poorly defined trail that is still strewn with winter debris. It is an old-growth forest like few I have ever seen. Thick stands of ancient Douglas fir and Western Red Cedar are surrounded by blankets of moss inches thick. I scramble, stymied by a massive log laying across the trail and continue on, huffing up switch-backless micro-slopes, gaining and losing the same one hundred feet of elevation over and over again. Twice I am lured by side trails that lead to research stations and am forced to double back, searching for the trail. Along the way I notice a tiny white flag, a marker for some kind of research, and I can't help feeling that forest is signaling surrender, aware of its tenuous relationship with these humans, the fragile network of policy and funding that keeps it from the chainsaws. Finally, I am stopped entirely by a tangle of fallen limbs too thick for me to pass. I retrace my steps back to my car.

I continue driving up the steep, rocky road to the other end of the trail. The road is narrow enough that even my tiny car has to thread between the cliffside and the giant potholes that pock it. I stay in first gear, collecting dust, eager to arrive intact, which I do. Back on the trail it's clear from the outset that few people make it this far into these woods. The path is faint. An eighth of a mile in, it's also snow-covered. I step through the first two patches using the prints of a larger foot, likely a ranger. But ahead I see a longer stretch of white and deep holes where my guide has fallen through to his calves. For the second time, I turn around.

MAY 3, 2019, DAY FIVE

THE MAP IS POORLY LABELED. TODAY I LOSE HALF AN HOUR TO confusion, walking mislabeled roads, roaming uphill and back again in search of entry to the forest. Everywhere trees lay piled like kindling from the storm, blocking my way. I climb over and scramble under them for an hour, only to finally be turned away by a trail so steep and perilous I doubt it should be hiked by anyone in solitude.

Defeated and walking back to headquarters I realize why this patch, this one percent, was gifted to the ecologists and researchers, why these ancient trees and these alone were offered up for protection at the height of Oregon's logging: The forest will not release them. The steep, impenetrable terrain paired with the soft, ever-failing volcanics make logging, with its heavy equipment, laughable. I see now that it is not the benevolence of humans that preserved this forest but the land itself. Our perceived roles as triumphant stewards and conservers of this forest are a construct, a fallacy. This is not a bastion but a reservation.

But then today, from my cabin with the windows open, the sound of children laughing is brought to me on the breeze. Their voices chuckle and roll like the stream over the rocks. They are girls, at play in nature, insulated for a time from the pressures and perils of modern girlhood by the canopy of these ancient trees.

I follow the lilt of their laughter into the woods. They are middle grade, and I am startled by the freshness of their bodies, the softness of their skin visible from even a distance. I close my eyes and for a moment see myself as a young woman in these woods, standing, much as they are now, circled around a group leader, smiling in the sun. I am told they are here as a coming-of-age experience, a rite of passage. They are all wearing Forest-Service issued helmets, a sign of our litigious times. I wonder how my early impressions of the woods might have been different if I was doomed to experience it in this aseptic and risk-averse way? Will they carry this practice into adulthood, the sound of the forest forever muffled by padding? After some introduction they are each given a location along the interpretive trail and asked to make their way on their own. Then, they sit. Or stand. Alone with their thoughts for a time in the woods.

Back in the Green House I picture these girls in their forest plots, and see them as the vine maple, appearing to always be lucky, catching what little light penetrates to the forest floor in their uplifted palms. Youth, it seems, is always gifted grace. In truth, it runs the other way, the most successful maples have rooted themselves within grasp of daily light and it has taken them years to grow into that space. The forest shines bright with their golden green because they have chosen it to be so. The girls, I know, must also choose to grow into the light.

Shortly, I hear the sound of their return down the dusty road. I am startled. In this era, an hour, apparently, or even less, is sufficient to affect one's passage into adulthood.

MAY 4, 2019, DAY SIX

I HAVE BEEN LIVING HERE DAYLIGHT TO DAYLIGHT, WAKING just before sunrise and collapsing into sleep before true darkness. But today I rise at three a.m. to watch the stars, the first of six short hikes I will take today, one approximately every two hours, set on a time-based experience of the Andrews. When I step outside in the darkest hour before dawn, I spin, reeling backwards as the field of stars expands my vision. We are passing through the tail of Halley's Comet and the normally crowded sky is promised to be filled with meteors.

I see more satellites than shooting stars. Human endeavor intruding on the natural world. I feel like I've been brought here not to observe nature but to observe man's touch on nature, the ways in which we are slowly seeping into even her most prohibitively guarded places.

At first light I set out again, returning to the road that began my time here. I search and crawl through understory until I find the way to the trail that runs along Lookout Creek, and beyond a tangle of slim, fallen firs I see an aged and weathered sign warning me that the bridge is closed three miles ahead. Pressing on, the trail disappears beneath the tinder around the second bend. Again, the Andrews turns me back.

Two hours later I emerge again, this time heading out through Headquarters and up a steeply inclined gated road that runs high above camp. It is always uphill in all directions in the Andrews. The chipmunks and I share the walk with a bejeweled bug I have never seen before. It paces me, slowly moving up the steep gravel path, a rectangular fluttering of red. I have had the solitude of early rising for five hours now. The sun has slid to the base of the trees across the road, but has not yet touched ground. The woodpeckers and crows have begun their racket, drowning out the lyrical songbirds of earliest morning.

It is chainsaw day. By midmorning engines fill the forest and cars and trucks start racing up the roads with none of the caution of the workdays. These are graduate students, I am sure, driving in from Eugene and Corvallis, slaves to the university during the week, destined to be short of time and energy by the weekend. I hear them call to each other through the trees, unaware of having company, enjoying filling the endless space around them with their voices.

Hours pass and the still chainsaws roar. They are a constant buzzing punctuated with deep, guttural roars as some piece of equipment or other devours a log. It is necessary work, I know, and I am glad for it, after so many days of being turned back by fallen trees, the risk of fire so clear upon the ground. What will come of this place, so dependent on long-term measurements and reflections, if it is all to burn away?

That, too, is a data point, I suppose.

By the afternoon I am on my fifth hike of the day and the air is hot and smells like dirt. The river is cacophonous, somehow louder than in morning, the few remaining birds barely audible above its din. Midday is pensive, not like the opening of morning, when everything seems in motion. Instead, everything seems to hold its breath in the heat of day, the plants are busily eating up the sun but everything else shies away. The chipper still roars, like a dying animal, choking on its kill. The machinery blocks the road; I couldn't leave if I wanted to.

In late afternoon my final walk provides relief. The machines are gone. Cool air slides down the valley walls. The birds are still and silent. A stick in the road transfigures before me into snake, sliding away in front of me just beyond my steps. The sun slants through the trees, reaching now to the forest floor. The world glows the soft green of moss and new growth. The river calms, allowing for the voice of a gurgling stream to be heard above its din. Everything dances in the breeze.

MAY 5, 2019, DAY SEVEN

THE DOGWOODS HAVE COME INTO FULL BLOOM, THEIR WHITE flowers like kisses on the pages of a letter. Today I take more pause, looking at the old-growth trees as a community, rather than individuals. I observe how they sit amongst each other, set at a deliberate distance, together, but apart. I wonder what it's like to stand next to another living thing for so long. How many generations of vine maple and squirrels has this community, this stand of trees, witnessed playing at their feet? What were they like in their adolescence? Did they let the wind take their boughs more freely? Did they dance roughly and mingle more with those around them before gaining their current stature?

The Andrews does not give up her secrets sweetly. I take the gated road again, determined to not be turned back. It is a climb, seemingly forever, uphill into mountains. There are trees and debris all along the road, which has a near-vertical slope to the side of it. I move slowly upwards, climbing over and under piles of debris and logs. I know that my exploration here will be penultimate. I will be allowed no summits or completions, but I press on.

Until at last, a view to the distance, a panorama of the entirety of the Andrews. It is the first since my arrival. I stand for several minutes at the edge of the road, my arms outstretched to the world, restoring my soul, finally feeling transformed by this place.

But then from the corner of my eye, a camera. Its gaze fixed directly over the place in which I stand. My arms drop to my sides. I know that cameras are ubiquitous in the Andrews, mounted on fence lines, buildings, equipment stands, and even the trees themselves. I want to not care. But still I am surprised. I take a picture of it taking pictures of me. I wonder how much of my time here has been recorded and what we are recording it all for. Surely, if a tree falls in this forest, someone hears it. Why can we not bring ourselves to trust that the sediment, rocks, tree rings and erosion are recording their history for us? If only we would learn to listen. I try to turn my gaze back to the forest, wanting to continue being in the world in such a naked and open way. But the camera's presence makes me lose the glory of the place, eventually urging me down the hill.

MAY 6, 2019, DAY EIGHT

IN EIGHT DAYS, I HAVE PASSED NO ONE IN THE WILD. NOR HAVE I encountered another human more than a few feet from either a vehicle or a building. The row of vehicles outside the living quarters and parked in the Headquarters' lot suggests there are at least a dozen people in residence, more than twice that are on site during working hours. But in off hours, I see no open windows, no one sits on porches or at the picnic tables. Lawn chairs are covered in pine cones and debris, suggesting it has been some time since their use. There is no gathering of souls, no signs of human life save construction, infrastructure. I try to imagine the scene as people lived even twenty years ago, lounging outside with a book, playing guitar or volleyball in the now abandoned court. Instead, I see it as studies tell us most Americans live, inside hunched over their phones playing Fortnite, streaming YouTube beauty videos, shopping on Amazon, refilling their prescriptions online, eating microwaved food.

And these are the ones that want to be outside.

Yesterday I felt a sadness about the brevity of my time here. Today, I feel an urge to leave. The Andrews for me has become a microcosm. It is a place where humanity presses itself into and against nature, where people work but do not revel, science and engineering reign supreme and ancient trees are drenched in wires—a digital forest. My time here, I know, is meant to be a part of this machine. I have been asked to reflect on a carefully curated part of the forest. None of the long-term reflection plots are in the clear-cuts.

I pack up and drive down the road throwing a cloud of dust into the air behind me, sure of only one thing from my time here: the Andrews is being studied, but we have not learned what it has to teach us.

IT'S ALL
GOING TO FADE

T IS A BRIGHT SPRING DAY IN 2019, A RARE BLUEBIRD MORN-
ing for this most-northern part of Washington. Paul and I
are driving south from a weekend in Bellingham, a port city
near the Canadian border. We have risen early and we have the
narrow, winding highway along Washington's jagged coastline
nearly to ourselves. We are heading home to Eugene, Oregon,
a trip that will take us four hundred miles through nearly the
entire length of the Pacific Northwest. It is a familiar drive along
which I can map the events of most of my life.

Space and distance are still abundant commodities in the
Pacific Northwest and there is plenty of road to put beneath us.
We drive. To the west is the Salish Sea, freckled with the San
Juan Islands, the long ridges of the Olympic Mountains just
visible beyond them. To the east, the horned peaks of the North
Cascades rise, evidence of the ongoing process of accretion in
the region. Farther south the road plunges towards sea level
and we enter the rich, glacially-derived agricultural flats of the
Skagit Valley, known for its tulips, strawberries, and rich coastal

fishing grounds. As the terrain flattens, the high Cascade volcanoes rise and sparkle in the sun like a string of diamonds above a vast blanket of deep-green trees. The miles pass beneath us. I track our progress by familiar forms, the jagged, snow-covered silhouette of Mount Baker, the squat Glacier Peak with its twin summits, and farther in the distance, the rounded tops of Mounts Rainier and Adams.

Too soon, the highway merges with the interstate and open country gives way to development. Now the scene is dominated by shipyards, military installations, and the sprawling campuses of tech-country. The view narrows into the familiar commercial monoculture of the I-5 corridor. Clusters of big box stores, many now defunct, line the freeway along with car dealerships, mattress outlets, and an equal measure of adult stores and family fun centers. Fast-food restaurants serve long lines of cars greasy breakfasts. Concrete walls barely obscure close-set ticky-tacky houses. For the next one hundred miles there is little that distinguishes this stretch of American road from any other. At Seattle, the traffic increases as life in the city unfolds in pretty much the same way it has since the turn of the century. Children stream videos on iPads as their parents stand on line in Starbucks scrolling through the day's headlines: the beginning of Gay Pride month, a worrisome measles outbreak, the #MeToo movement, and a Federal investigation into subpar practices at Boeing among them. The Space Needle slips by, dwarfed by glassy high-rises.

South of Puget Sound we plan to break away for a lengthy detour to a place that's easy to bypass in spite of the way it dominates the now-open view. Mount. St. Helens, with its gnarled half-top and horseshoe shaped crater sitting open-jawed and gaping in the distance, is plainly visible from Seattle to Portland and not on the way to anything. A visit to Johnston Ridge will take us nearly two hours from the freeway. But Paul has never been, and lately it has been very much on my mind.

You could say the course of my life was determined by that volcano, and for a long time that was the story that I told myself. That its eruption shaped the landscape to which I was born is certainly true, as is the fact that the opportunity to study its inner workings spurred my career as a geologist. But recently, I have begun to question that narrative.

Still, that mountain, and the events of May 18, 1980 in particular, loom large in my imagination, though I have only one memory of that day that I can claim to be my own. In it, my mother stands in our front yard holding me in her arms and watches the plume rise into the air, darkening an otherwise clear spring sky. Other memories of the volcano are secondhand but vivid, unique in their sensory nature. In one many years later, I sit in a darkened high school classroom watching footage of twenty-eight-year-old news reporter Dave Crockett run from the eruption. As the film begins, he is casual, taking in the entirety of the scene and training his lens on the rising plume. Within seconds though, the plume has expanded dramatically towards him and as the sky turns nearly black, he begins to run. There is little to see in his jostling footage, but the deep rumble of the mountain, the crush of his feet across the rocks, and his heavy breathing are clear. The projector clicks and flips away, adding an ominous sense of realness to the scene. Breathless, I can still hear him say, "At this moment I honest to god believe that I am dead." Or later, my years of undergraduate research meld together into a repetitive stream of study halls, lab work, and data points. But I can still feel the weight of U.S. Geological Survey Professional Paper 1250, an 844-page tome produced in the wake of the eruption that twenty years later still constituted the majority of our understanding of that system and that I was rarely seen without in my four years at university. It still sits on my shelf today.

The landscape unleashes another flood of memories. In September 2004, after years of calm, a swarm of near-surface earthquakes known as harmonic tremors announced the arrival of fresh magma to the system. By October, small-scale steam and ash eruptions had begun to rise above the rim and a spine of hot rock emerged like a breaching whale in the crater. As one of the few people still actively researching her, I was asked to help with the monitoring effort. I had just completed graduate school and was still finding my way, just months into my first job as an environmental and hazards consultant and teaching geology at a local community college in the evenings and on weekends. The work on Mount St. Helens was unpaid, but I said yes anyway, hopeful that this eruptive sequence would be my entry to a career in volcanology as it had been for the young men and women that monitored the eruption of 1980, the same people who still held those positions two decades later.

It would be the most intimate period of my relationship with that volcano, marked by long hours of driving from the observatory in Vancouver to ash-collection points inside the newly established evacuation zone. Unlike in 1980, the public in 2004 heeded warnings of the dangerous and volatile potential of the volcano, leaving it entirely to researchers. On one such excursion, I was accompanied by a PhD student from Oregon State. Just before sunset we opened a road gate and entered the evacuation zone, locking it behind us and made our way to the farthest collection station, a plastic bucket secured to a pole just short of the ridge where Dave Johnston had been stationed. It is eerily quiet and devoid of people. After collecting the traces of ash from the previous day's small eruption we stood in the middle of the road, alone for miles, and watched a column of steam and ash rise from the newly formed dome. Just days earlier I had been prepping for a Geology of the Pacific Northwest class at a community college outside Portland when two of my students came running in, breathless. The mountain was erupting, the plume was clearly visible from the balcony of the building. I held class outside that day, collecting the instructors and students from the surrounding poetry and humanities classes, all of us thrilled to watch what for most of us was our first experience of a volcanic eruption in real time from the safe a distance of over a hundred miles. Days later, standing in the shadow of the gently erupting giant at sunset, only a handful of miles from the growing dome, it was easy to forget the mountain's dangerous potential. But we both knew the most predictable thing about these kinds of volcanoes was their unpredictability. After several minutes my companion gently nudged me, reminding me of our isolated location within the evacuation zone, the locked forest road blocking our exit, and the potential, at any moment, of an increase in eruption intensity. Reluctantly, I headed back to safety, my eruption fever far from quenched.

Several weeks later I got one more chance to play with the volcano's fire, flying directly through the erupting plume as part of a three-person gas-data collection crew in a small plane manned not by Forest Service pilots who had deemed the trip too dangerous but by John Pallister, a researching volcanologist, one of the original "volcano cowboys" from the eighties, who had built the plane himself years before for just that purpose. I can still feel the way my stomach lurched and sank as we passed through the eruption plume, the plane dropping several hundred feet in response to the dramatic change in air pressure and temperature. "Just a little bump," John said with a chuckle.

Over the years I have kept these stories in the pockets of my mind like worry stones, working and reworking them until that devastated terrain, that place of ash and heat and failure, became welded to my personal folklore as the pivotal landscape of my life. Now, even after nearly a decade away, I feel its pull on me again. And like a salmon to its headwaters, I return.

South of Puget Sound we enter the Chehalis River Valley, a broad, irregularly shaped watershed that is home to an unusually circuitous and slow-moving river. Gathering water from over two thousand square miles of land, the Olympic Mountains and Puget Sound to the north, Cascade foothills to the east, and the Willapa Hills to the south, the Chehalis Valley is the largest drainage basin in Washington. It is a confluence of water, rock, and culture. Here, ancient seafloor is overlain by the northern lobes of the Columbia River flood basalts, glacial till from the last ice age, and young volcanic deposits from the Cascade volcanoes. Atop these deposits is a convergence of infrastructure—the roadways, bridges, and rail lines that transport raw materials and resources for the timber and fishing industries, and, more recently, the hotels, casinos, and development of the Confederated Tribes of the Chehalis Reservation, who have become one of the region's largest employers. From the interstate, the valley is a string of hard-luck western towns surrounded by timberland, the most memorable landmark being a fifty-year-old billboard depicting Uncle Sam next to a constantly changing series of alt-right slogans.

"I'LL TAKE GOD, GOLD, AND GUNS. YOU KEEP THE CHANGE."

We drive on.

In the truck, I am weaving my volcanic folktale, the words spilling out fast. "You have to see this place to believe it. It's like a Martian landscape. Everything is gray and nothing grows, it's all been choked out by piles and piles of ash. Once it got started, eruption went on for more than eight hours. The landslide released all the pressure from the bulge like a pop can. Half the mountain slid away and released the magma which melted the glaciers so the rivers all flooded with this concrete-like mixture of super-heated water, pumice, and ash from the pyroclastic flows, and downed trees from the blast. It was so powerful that the landslide rode up and over ridges, that's how that old boy Harry Truman who refused to leave his cabin died up at Spirit Lake, the whole mountain just rode over the ridge and buried him. The pyroclastic flows traveled down the river valleys, incorporating the water and turning into volcanic mudflows called lahars that traveled up to 150 feet per second, scouring the land in some places, burying it in others. There's still feet and feet of it all along the banks of the Lewis and Toutle rivers. But my favorite part is the matchstick trees. When the lateral blast happened, it sent a shockwave through the air that knocked down trees across a huge area around Johnston Ridge. I mean, centuries-old trees were just snapped in half. And they had all their limbs and bark blown right off. It's like a bomb went off. There's just thousands of burnt trees and shattered stumps laying everywhere for miles. In some places they're blown down in circles and spirals, still preserving the swirling, internal flow patterns of the turbulent air from the blast. And it's not just the eruption that makes the place spooky. Most of this land is owned by logging companies; outside the blast zone, the rest of the forest has all been clear-cut or salvage logged."

Like many residents of the Pacific Northwest, I can see all these things—the matchstick trees, the eruption column, the white-gray of the pumice plain, the Toutle river choked with ash—clearly in my mind's eye. The eruption in 1980, with its slow onset and close proximity to urban population centers, was one of the most intensely studied geologic phenomena in the world. Every steam blast and harmonic tremor was fodder for evening news eager to satisfy the growing appetites of a newly hyper-connected public. And, unlike nearly every other geologic event, everything at Mount St. Helens was documented in real time, making it possible for aspiring scientists such as myself twenty years later to become experts on events that occurred while our parents still held us in their arms.

And while I have spent my share of time on her flanks, in truth, almost everything I have ever known about Mount. St. Helens I learned from a great distance of both time and space, or to put it in geologic terms, through *remote sensing*. Remote sensing is a catch-all phrase referring to the acquisition of data without close contact: information gathered from aerial photographs, satellite images, sonar, ground penetrating radar, seismic data, ultrasound, lasers, and heat measurements. Geologist use these techniques more than most other sciences. They are tools used to collect information about things too deep, hot, inaccessible, or volatile to examine in close proximity. The data collected from these sources is secondhand, based on the interpretation of a returned signal of light, sound, or energy that is sent out from a source and relayed back. The returned signal is then processed, producing images, data sets of arrival times, temperatures, or sounds, all of which must then be interpreted. Circles and lines in aerial photographs become roads and lakes, spikes along a seismograph are interpreted as shallow earthquakes or herds of mountain goats across the ground. It requires both imagination and precision and is as much an art as it is a science. Often, it is an imperfect process. Remote sensing then requires returning to an area to verify that the interpretation of the data is correct. The findings are called *the ground truth*.

History, especially personal histories, are a form of remote sensing, an interpretation of ourselves from some distance in time or space. We base our identity, our sense of self, on memories, which in truth are little more than echoes, signals that are reflected and recovered but changed somehow in the return. Still, viewing our lives in retrospect has both purpose and allure. But like remote sensing, it is subject to interpretation, to the evaluation of unreliable data, hazy memories, misinterpretations, misunderstandings, and lost signals. At midlife, where I find myself now, these uncertain signals have produced a deep craving to find the ground truth of my experience.

Just before noon we turn off the freeway and head east, towards Mount St. Helens for the first time in nearly a decade.

The road meanders through a verdant agricultural valley framed by cascade peaks. It is lonely, even though it is a sunny weekend day. Eventually we pass a small town, little more than a cluster of buildings at a crossroads. Tattered and faded signs advertise gas, bait, and tackle, and Mount St. Helens tchotchkes, but none of the handful of businesses are open, not even the gas station. In 1980, before the eruption, this place would have been teeming with fishermen towing boats to mountain lakes, pickups filled with raucous teenagers, and families heading out for weekend camping trips. Even at the turn of the twentieth century the narrow road would have been clogged with volcano tourists and the local store would have made fast money on deli sandwiches made with white bread and American cheese and dime-bags filled with Mount St. Helens ash. The quietude is disconcerting, my first indication that something is not right. "Where have all the people gone?" I ask Paul. People don't go outside anymore, he reminds me, and no one cares about this volcano anymore; no one remembers.

By the time we reach Spirit Lake Highway, the road to Johnston Ridge, confusion begins to set in. It is in many ways exactly as I recall from my most recent visits leading gaggles of students on field trips. It is still speckled with pullouts to vistas, overbuilt, expensive-looking visitor's centers, and massive parking lots that announce easy day hikes and clean bathrooms, but the road is open, the parking ample, and the vistas are blocked by stands of conifers. Disoriented, I miss one of my favorite stops, a road-cut exposure of the deposits showing the internal patterns of the pyroclastic flows with a view across the pumice plain and down to the Toutle River. As we pass, I realize that in the twenty years since I first came to this place the trees have grown, obscuring the once clear view. At the viewpoint of a newly constructed visitor's center I peer through a viewfinder at the pumice plain below. In forty years the river has shifted the ash, forming an intricate network of incised channels and braided streams that weave around rocky hummocks, house-sized chunks of the north flank transported downstream by the landslide. The once gray-brown surface glows a gentle green in the bright spring sun and dense stands of brush encroach from either side, contracting its reach. Mount St. Helens is in front of us, revealing little of her collapse from this western approach save her flat top. Next to me, Paul is underwhelmed. And rightly so, after my chronicle of a Martian landscape. Instead, he comments on the area's bizarre beauty, pointing to broad swaths of replanted noble fir forest planted in such uniform monoculture that from a distance their limbs produce a pixelized effect.

It is a remarkable transformation and one that has thus far run entirely against prediction. In the fall of 1980, just months after the eruption, scientists proclaimed that the area in the immediate wake of the event would remain barren and lifeless, perhaps for centuries. The force of the blast, the depth of pumice and ash deposits, the heat, and the lack of nutrients, scientists said, had created permanent damage on a human timescale.

But only a handful of years later, it became clear that Mount St. Helens would, in fact, recover. And that recovery, the return of a healthy and abundant ecosystem, would be faster than they ever imagined possible. Twenty years later, when I began leading field trips, the embryonic new forest was already being heralded as remarkable by biologists and ecologists. My geologist's eyes saw only the pumice plain, the cuspate failure of the peak, the swirls of downed trees. I was blind to the saplings, birthed from seed carried in by the wind and on the backs of birds and chipmunks, that dotted the surface of the flow. I took for granted that Spirit Lake, once buried along with Harry Truman, had reestablished its surface waters, and I had scorned the green and white signs that lined the road proclaiming the year each clear-cut had been replanted.

Ironically, in the weeks before this trip, Fred Swanson, my guide in the Andrews forest, upon hearing of my previous work on Mount St. Helens, had sent me his seminal work. It was a review of the research to date concerning the recovery, which had by then informed a whole new field of study established in response to the effects of human resource use, population growth, and influence on climate-disturbance ecology. What they found to be true at Mount. St. Helens (and then elsewhere) is that recovery progresses in ecosystems in much the same way that it does humans, through a combination of physiology, site amelioration, community assembly, succession, and biotic inter-actions. That is to say, the progress of recovery, the reinvention and rejuvenation of the self, depends on a complicated inter-play between our bodies, communities, and environment. It is a process whose progress is deeply influenced by the random nature of the universe. Ultimately the researchers concluded that however impeded or unlikely recovery may seem, even-tually, it comes.

I knew all of this. But it did nothing to prepare me for the shock of arriving to this once familiar place to find it so transformed. By the time we arrive at Johnston Ridge, I am just another tourist. I wander through the gift shop, pause to take in the interpretive displays, and stoop to squint through a volunteer's scope at the crater, the whale-backed dome still visible in its center. From this perspective, the enormity of the catastrophe is still well preserved. I listen as the volunteer enthusiastically describes the events of May 18, 1980, and look with interest as she trains the lens on relevant features of the scene before us—here is the crest of the broken rim, the lumpy surface of the landslide, the ash deposits from the plume.

Finally, we break away, moving along a narrow trail that eventually leads to the mountain's summit, or what passes for a summit. The ground is composed of compressed ash and pumice that crunches under our feet. In defiance of the lack of topsoil a spray of wildflowers covers the surface in streaks of red and purple as far as I can see in all directions. At the crater, a dusting of late-spring snow covers the upper third of the peak, an unusual and refreshing sight in these climate-heated times. I stand on a narrow ridge squinting into the reflected sunlight, try to calibrate my thoughts, and search for the truth. For a moment I am unable to categorize the land around me. Is it a place devastated or rejuvenated?

The new dome, destined to someday reform Mount St. Helens peak, releases a ribbon of steam into the still-clear blue sky. I stand facing her and understand that I am at once an expert on this place and a dilettante. My expertise and that of every other scientist is only a mere slice of a history that began when the current configuration of the Pacific Northwest began to override the Pacific plate, forcing it to plunge beneath the continent beginning the process of subduction, melting, and magmatic rise that would eventually form the Cascade volcanic arc. The eruption of 1980, while defining the mountain to its twentieth-century human observers, was only one event in the forty thousand-year lifespan of a mountain that has in that time evolved from a fissured opening to a towering composite volcano, seen the advance and retreat of ice sheets, witnessed the passing of Halley's Comet more than five hundred times, and built and wasted and built itself over and over again. From that wide-angled perspective it is equally defined by the planting of a single seed dropped from a chipmunk's back as it is from the failure of its north flank and no amount of study or documentation of that day will ever characterize it completely or truthfully. For the first time, I see it as a living, respirating entity that exists only in the present, continuously reinventing itself in response to forces beyond its control and the random nature of the universe. On this day it has chosen to be magnificent.

Mount St. Helens disappears from the review mirror quickly on the drive home and finally we are in more familiar territory. Late in the afternoon we cross the Columbia River, passing the turnoff to the long-demolished Cleveland House and the dam with its counting window beyond it, into the sprawling development of the Portland area and past the base of Council Crest. The clear skies have held and I know that today it will bustle with Portlanders eager for a view of the mountains, Mount St. Helens among them. We enter the flatlands of the Willamette Valley bright with the rebirth of spring rooted in glacial outwash. Fields of white meadow foam, crimson clover, and variegated tulips shine against an infinity of blueness punctuated by undulating ridges of Douglas fir–forested hills.

The truck drives on and I float above the road, a creature at once the same and entirely reborn. I am ready for a life unencumbered by the demons, memories, events, and people of my past, sure of who I am and unafraid of judgment. In the bedrock, water, forests, and deserts of this region I have at last found my groundtruth, which as a single word is defined simply as a "fundamental truth." For what we look for when we search for truth of the self is not scientific certainty or some kind of divine knowledge but authenticity. What we long for is to know and accept ourselves sincerely, as a work of art in the raw and *sin cera*, "without wax" to cover our imperfections.

I feel no disappointment from being unable to show Paul the devastation at Mount St. Helens and no need to continue to return to the devastated terrains of my life, to ground truth the lost souls, bad relationships, missteps, or lessons learned along the way. Instead, I want to ask them the questions that have brought me to this moment. "What is the landscape of your life, a wasteland, or a paradise? Where are you on your journey, your way out or back in? What have you become, a junkie or a chef? And what shall you do now, live a life of your choosing or be buried in the ash?"

Because we live in the moment. This magnificent moment. And no amount of retracing our steps or recounting our history will change that. It might help illuminate the path we took to get here, but the future remains for us to step into, again, and again, and again, secure in the knowledge that what comes to pass in our lives, no matter the immediate impact, will surely fade over time and reshape us into something else, like a vibrant spring day on a still-steaming volcano.

ACKNOWLEDGMENTS

'D LIKE TO EXPRESS MY GRATITUDE TO THE PEOPLE, INSTITU-
tions, and publications that informed, inspired, and
supported this project over the years; this was by no means
a solitary endeavor.

To my wise women and first readers Patricia McConnell
and Morgan Devos-Rosemarin I extend my eternal love and
gratitude. Thank you for guiding me. To my father, James
McConnell, whose dedication brought me into reading and
whose hand will always hold me safe, thank you.

To the many teachers and mentors that have educated,
encouraged, and challenged me over the years: Dana Heffner
Wallmark and Nancy Lapotin from Lincoln High School,
Kathy Cashman, Marli Miller, Becky Dorsey, and Dana
Johnston from the University of Oregon, Nancy Riggs,
James Wittke, and Roderick Parnell from Northern Arizona
University, Gerardo Carrasco-Nunez from the University of
Mexico Queretaro, Carl Thornber and John Pallister from
the USGS, Fred Swanson from the H.J. Andrews, and the
one and only Martin Acaster, thank you. Good educators
change lives.

I am eternally grateful to the many writers and editors and friends that helped me find and amplify my voice with this project: Hannah Elnan and Gary Luke of Sasquatch Books, Ann Wiens, who was the first to publish my work, Ben Waterhouse and Kathleen Holt of Oregon Humanities, Noah Michelson, whose gift of time and compassion made the hardest of these essays possible, Ann Littlewood for her mentorship and guidance and Daniel Littlewood for being a true friend and sharing with me a very important mango, Jo Robinson, who helped craft the proposal for this work, Michael Copperman, who elevates all Oregon writers, and others including Amy Lokach, Denise Vilanova, Kyle Strauss, Julia Park Tracy, Kim Stafford, Ellen Waterston, Jonathon Evison, Jeff Geiger, Eric Alan, Mary Democker, and Scott Lanfield. This would not have been possible without you.

Thanks also to the organizations that provided essential funding in support of this project and the publications that printed early versions of these essays: Oregon Literary Arts, the Spring Creek Project, Oregon Quarterly and the Northwest Perspectives Essay Contest, Willamette Writers Kay Snow Contest, Oregon Humanities, *Grain Magazine, Huffington Post, Burnt Pine Magazine, Entropy* magazine, and Our Bodies Our Voices women's reading series.

To Patrick McDonald and Rachel Bell of Overcup Press, I cannot express the profound nature of my gratitude to you for believing in this project and shepherding it to fruition. A thousand thanks to you.

To all those people who studied, explored, and adventured with me all these years, far too many of you to name—stay wild my friends.

And finally, to Paul Hampshire, my love, my prince, my heart and soul. Without you there is nothing.

SELECTED BIBLIOGRAPHY

Abbott, Adler, and Post Abbott, *Planning a New West: The Columbia River Gorge National Scenic Area*, Oregon State University Press, 1997

Alden and Paulson, *National Audubon Society Field Guide to the Pacific Northwest*, 1st Edition, Knopf, 1998

Alt, David, *Glacial Lake Missoula and Its Humongous Floods*, Mountain Press Publishing Company, 2001

Alt and Hyndman, *Northwest Exposures—A Geologic Story of the Northwest*, Mountain Press Publishing Company, 1995

Alt and Hyndman, *Roadside Geology of Oregon*, 1st Edition, Mountain Press Publishing Company, 1978

Ambrose, Steven, *Undaunted Courage: Meriwether Lewis, Thomas Jefferson, and the Opening of the American West Paperback*, 1st Edition, Simon and Schuster, 1997

Arrowsmith, Robin, *All the Way to the USA: Australian WW II War Brides*, 2013

Attwell, *Columbia River Gorge History*, Vols. 1 and 2, Tahlkie Books, 1974 and 1976

Bates and Jackson, *Dictionary of Geologic Terms*, Anchor Books, 1984

Breitenbush Community Online History, Breitenbush.com

Carstensen, Armstrong, and O'Clair, *The Nature of Southeast Alaska: A Guide to Plants, Animals, and Habitats (Alaska Geographic)*, 3rd Edition, Alaska Northwest Books, 2014

Cashman and McConnell, "Multiple Levels of Magma Storage During the 1980 Summer Eruptions of Mount St. Helens, Washington," *Bulletin of Volcanology*, 2005

Chehalis Land Trust Online Resources, www.chehalislandtrust.org

City of Portland, Oregon Digital Archives, www.portlandoregon.gov

City of Portland, Oregon Parks Department Online Resources, portlandoregon.gov/parks

City of Rockaway Beach, Oregon Online Resources, corb.us

City of Seattle, Washington Digital Archives, www.seattle.gov

City of Seward, Alaska Chamber of Commerce, www.seward.com

City of Yachats, Oregon Online Resources, yachatsoregon.org

Clark, Ella E., *Indian Legends of the Pacific Northwest*, University of California Press, 1953

Cook, Scott, *Curious Gorge—Hiking and Exploring the Columbia River Gorge Area*, 4th Edition, 2017

Finn, J.D., John, *Heroes and Rascals of Old Oregon: Offbeat Oregon History*, Vol. 1, Ouragan House Publishers, 2019

Friends of Vista House, vistahouse.com

Gulick, Bill, *Roadside History of Oregon*, Mountain Press Publishing Company, 1991

Harney County Library Oral History Project, harneycountylibrary.org/oral-history-collection.html

Harney County Oregon Online Resources, www.co.harney.or.us

Hilbruner, Van Camp, et al., *Columbia River Gorge: The Story Behind the Scenery*, KC Publications, 1995

Hogstad, Jason, "War on Rabbits Begins Sunday: Ritual Rabbit Slaughter and the Oregon Extension Service," Oregon State

University Online Resources, blogs.oregonstate.edu, October 24, 2016

Hylton, Hilary, "How Rabbits Can Save the World (It Ain't Pretty)," *Time* magazine, December 14, 2012

Idaho Interior Department, "Craters of the Moon: A Guide to Craters of the Moon National Monument," 2010

Kenai Fjords National Parks Service Online Resources, www.nps.gov/kefj

Kern County Generations Online Resources, kerncountygenerations.org

Kyselka and Lanterman, *North Star to Southern Cross*, The University Press of Hawaii, Honolulu, 1976

Library Ireland Online Resources, libraryireland.com

Lipman and Mullineaux (eds.), "The 1980 eruptions of Mount St. Helens, Washington, United States Geological Survey Professional Paper 1250," 1981

Litton, Evie, *Hiking Hot Springs in the Pacific Northwest*, 4th Edition, Falcon Guides, 2005

Lopez, Barry, "A Presentation of Whales," *Harper's* magazine, March 1980

Magnolia Historical Society, magnoliahistoricalsociety.org

Malheur National Wildlife Refuge Online Resources, www.fws.gov/refuge/malheur/

Mead, George, R., "Yachats: The Town Called 'Dark Water at the Foot of the Mountains' " E-Cat Worlds Press, 2012

Mershon, Clarence E., *The Columbia River Highway, From the Sea to the Wheat Fields of Eastern Oregon, 1913–1928,* Guardian Peaks Enterprises, 2006

Miller and Cowan, *Roadside Geology of Washington*, 2nd Edition, Mountain Press Publishing Company, 2017

NASA Climate and Sea-level Resources, climate.nasa.gov

NASA Earth Observatory, earthobservatory.nasa.gov

NASA Image Database, nasa.gov>multimedia

Nelson, Clifford D., "The Guestbooks of Crown Point Chalet (1915–1927): A Research Project," 2001

Nisbet, Jack, *Ancient Places: People and the Landscape in the Emerging Northwest*, Reprint Edition, Sasquatch Books, 2016

O'Donohue, John, *Anam Cara—A Book of Celtic Wisdom*, Harper Collins, 1998

The Oregon Encyclopedia, oregonencyclopedia.org

"Oregon Field Guide, Sneaker Waves," Season 25, Episode 2502, October 17, 2013

The Oregon Historical Society Digital History Projects, ohs.org/education/digital-history-projects.cfm

Oregon Secretary of State, Oregon Blue Book, 1950 to 2019

Oregon Secretary of State Archives, sos.oregon.gov>archives

Oregon State Game Commission, *Oregon Sportsman*, Vols. 3 and 4, 1915

Oregon State Parks and Recreation, Oregon Beaches—A Birthright Preserved, 1977

Oregon State Parks Online Resources, oregonstateparks.org

Oregon State University Image Archive, scarc.library.oregonstate.edu

The *Oregonian* Digital Archives 1990–2019, www.oregonlive.com>archives

Orr and Orr, *Geology of the Pacific Northwest*, 2nd Edition, Waveland PR Inc., 2006

Overton and McManus, *Saving an American Treasure: The Story of Vista House*, CreateSpace, 2014

Pfeiffenberger, Jim, The Complete Guide to Kenai Fjords National Park, Greatland Graphics/Puffin Pr., 1995

The *Portland Tribune* Digital Archives, pamplinmedia.com

Portland Women's Forum Online Resources, portlandwomensforum.com

Prince and Clark, *Portland's Goose Hollow* (Images of America: Oregon), Arcadia Publishing, 2011

The *Register-Guard* Digital Archives, Eugene Public Library microfilm, 2000–2019

Richardson, Dan, "Friends Say View Could Rival Crown Point," *Oregon Historical Quarterly*, Vol. 19, 2006

Romano, Craig, *Day Hiking the Columbia River Gorge: National Scenic Area, Silver Star Scenic Area, Portland-Vancouver to The Dalles*, Mountaineers Books, 2011

Rombauer and Rombauer, *The Joy of Cooking*, Bobbs-Merrill Company, 1953

Sale, Roger, *Seattle Past to Present*, University of Washington Press, 1976

The *Seattle Times* Digital Archives and University of Oregon microfiche, 1980–2019

Smith, William, *Irish Priests in the United States: A Vanishing Subculture*, University of America Press, 2004

Soennichsen, Joel, *Washington's Channeled Scablands Guide: Explore and Recreate along the Ice Age Floods National Geologic Trail*, Mountaineer Books, 2012

State of Washington Forecasting and Research Division of the Office of Financial Management, "Population Trends for the State of Washington 2017," 2017

Swanson and Crisafulli, "Volcano Ecology: State of the Field and Contributions of Mount St. Helens Research," in Crisafulli and Dale (eds) *Ecological Responses at Mount St. Helens: Revisited thirty-five years after the 1980 Eruption*, Springer, 2018

Thompson, Richard, *Portland's Streetcars*, 1st Edition, Arcadia Publishing, 2006

Thornber, Pallister, Rowe, McConnell, Catalog of 2004–2007 Mount St. Helens Dome Samples United States Geological Society Open File Report 08, 2008

Tillamook County, Oregon Online Resources, www.co.tillamook.or.us/

Toedtemeier, Terry, *Wild Beauty: Photography of the Columbia River Gorge*, 1st printing Edition 1867–1957, Oregon State University Press, 2008

United States Army Corps of Engineers, Engineer District, Portland, "1962 Bonneville Dam Pamphlet," 1962

"United States Forest Service, Resurrection Pass Pamphlet," undated

United States Geological Survey Cascade Volcano Observatory Online Resources, volcanoes.usgs.gov/observatories/cvo

United States Geological Survey Educational Resources, usgs.gov

Washington State Department of Fish and Wildlife Online Resources, wdfa.wa.gov

White, Craig, Geology of the Breitenbush Hot Springs Quadrangle, Oregon, Special Paper, State of Oregon, Department of Geology and Mineral Industries, 1980

Whitney, et al., National Audubon Society Nature Guides, Western Forests, Alfred A. Knopf, New York, 1985

Williamson and McCail, *Tales of the Seal People: Scottish Folk Tales*, Interlink Publishing Group, 1998

Willingham, William F. Water and Power in the "Wilderness": The History of Bonneville Lock and Dam, Revised Edition, U.S. Army Corps of Engineers, Portland District, 1997

Willis, Peg, *Building the Columbia River Highway: They Said It Couldn't Be Done*, The History Press, 2014

Wilson, F.H., Hults, C.P., Labay, K.A., and Shew, N., digital files preparation, and Wilson, F.H., and Hults, C.P., geologic map compilers, 2008, Digital data for the reconnaissance geologic map for Prince William Sound and the Kenai Peninsula, Alaska: U.S. Geological Survey Open-File Report 2008–1002

Zielinski, Sarah, The Colorado River Runs Dry, in Smithsonian Magazine, October 2010